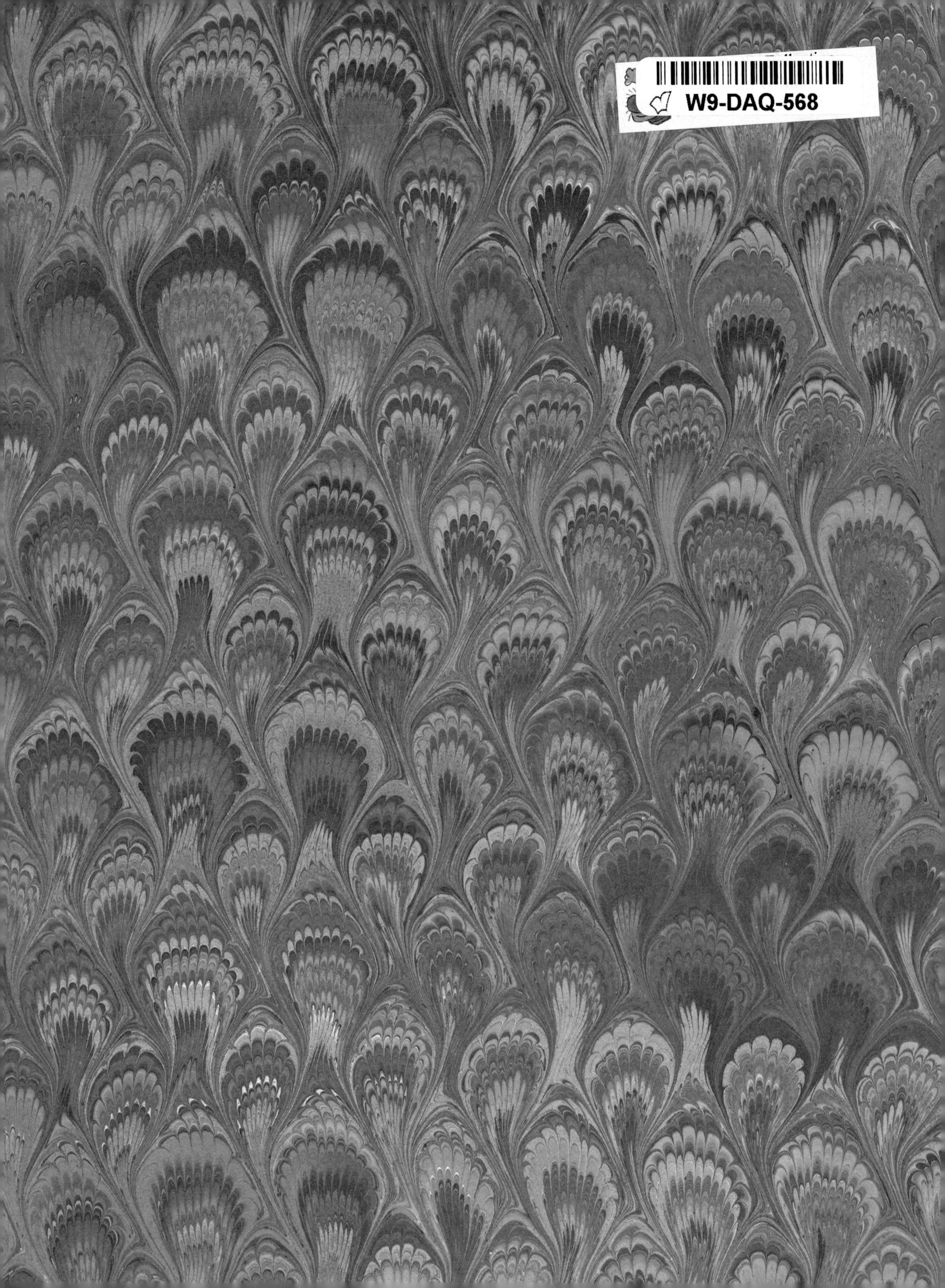
W9-DAQ-568

Beck's Superior Hybrids is pleased to present this limited edition volume, *Barns of Indiana* by Donald H. Scott. Proceeds from this beautifully illustrated book will go to help support the Purdue Agricultural Alumni Association.

Barns

of Indiana

By Donald H. Scott

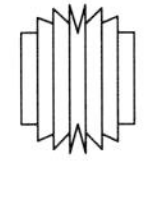

THE
DONNING COMPANY
PUBLISHERS

To Jackie, Pam, Jeff, Jim, and Patti

Second printing 1998

For information, write:

The Donning Company/Publishers
184 Business Park Drive, Suite 106
Virginia Beach, VA 23462

Steve Mull, General Manager
Ed Williams, Project Director
Paula A. Ridge, Project Research Coordinator
Dawn V. Kofroth, Assistant General Manager
Richard A. Horwege, Senior Editor
Chris Decker, Art Director
Tony Lillis, Director of Marketing
Teri S. Arnold, Marketing Coordinator

Cataloging-in-Publication Data:

Scott, Donald Howard, 1934–
Barns of Indiana / by Donald H. Scott : foreword by Maurice L. Williamson.
p. cm.
Includes bibliographical references and index.
ISBN 0-89865-995-7 (hardcover : alk. paper)
1. Barns—Indiana—Pictorial works. 2. Barns—Indiana—History.
I. Title.
NA8230.S39 1997
728'.922'09772—dc21 97–10462
CIP

Printed in the United States of America

contents

foreword

BY MAURICE L. WILLIAMSON

Historians, romanticists, visionaries, and other chroniclers of America continually search for a symbol that will somehow describe the vibrant, restless, entrepreneurial men and women who have, since the beginning, forged the destiny of this new nation.

Some would choose Ellis Island as it welcomed the hoards of immigrants who arrived to seek a new life. Others would choose the Statue of Liberty that seems to say, "Give me your tired, your poor. . . ." The sleek clipper ships as they plied the seas of the world certainly said, "I am America, and I am strong." Or, how about the pioneers and their billowing prairie schooners as they doggedly made their way westward, looking for a better piece of land to farm. Carl Sandburg probably would nominate Chicago with its skyscrapers, stockyards, and rollicking industry.

I will propose to you a humble but nonetheless magnificent symbol of our land—the American barn. Found almost everywhere you look, barns pay homage to the pride, the practicality, the ingenuity, and the enduring hope for the future of the people who understood the productivity of the soil, and worked diligently to harness those riches. Am I being presumptuous by suggesting that the barn be the chronicler of American history?

In the beginning, barns reflected little about the men who built them. They were born of necessity, and were little more than holes in the ground, or rude four-sided structures. As a matter of fact, they didn't even have a name. In Old English terms, *bere* ("barley") and *aern* ("place") became *barn*, ("a place for barley"). The Colonial barns were places to store grain and provender with no thought being given to housing livestock.

As communities became better organized and ethnic and religious societies settled in particular localities, their barns began to reflect their attitude toward beauty, pride, and utility, and became a substantial expression of their Old World origin. The thatch, the timber, the graceful stone clearly made the statement, "I am English, I am Irish, or I am Scotch." The great brick barns of the Middle Atlantic States were the transplanted attitudes of the Dutch and German ancestry. Barns built by Scandinavians, Swiss, Eastern Europeans, and many others all identified ancestry in the purest sense.

As time passed however, barns became more American and reflected the availability of indigenous materials, particular farming practices to be served, and were, most likely, a direct expression of the skills and tastes of the local barn builders.

Barns were very regional in character, and were built to meet the environmental and economic conditions of the area. The barns on the small Southern Appalachian farms were generally log or rough-sawn lumber and were never painted. The barns of the

East, the Great Lakes Region, and the Eastern Cornbelt were of pin frame construction, and utilized the trees of the virgin forests. Hand-hewn timbers of walnut, beech, and poplar sixty feet long were common, and stand to this day. Prairie barns were built of sod and cottonwood poles. Good timber was not available. Southern tobacco barns had a style all of their own with many louvers and cupolas. They generally were painted black, or not at all.

The colors of old barns remains one of the real folk mysteries of our time. Why were some red, others white? Why were some communities predominated by red barns and others by the white ones? Or do you remember that washed-out orange color that was called rather appropriately "turkey red"? And then there were the hex signs that were thought to be some sort of mystic mark of religious groups. (They weren't. They were just pretty and looked good on a barn.) The first color to be used to any great degree was red, primarily because materials were available to make the paint. Skimmed milk, lime, and linseed oil were the base ingredients. Grays of various shades used lampblack. The reds were made with ochre, red clay, or iron oxide, not from the blood of Indians as Colonial folklore intimated. The pale orange of turkey red paint did indeed come from the clay and the blood of turkeys. The fact that barn colors seemed to be predominant by communities was not necessarily a religious or ethnic thing, but probably reflected the wishes of the builders or the owners with others up and down the road following suit. In actuality, each farmer wanted to "outdo" their neighbors. Seemingly, the size, color, and the originality of the barn was a mark of superiority for the farmer. Bigger and brighter was better, they thought.

Nowadays, modern pole barns are tan, or light green, or mauve. Somehow they don't look quite right.

Those old barns were as enduring as the sturdy farmers who built them. With their slate roofs bearing the names of the owner, tall and stately cupolas, neat overhangs for sheltering the manure spreader, and the stout foundations, they were built to last forever. There was never a thought that a barn might outlive its usefulness and be abandoned to the cruel ravages of wind and neglect, to become leaning monuments of a long past glory.

The great American barn, whose time has come and gone, is now, happily, the focal point of frantic activity by preservation groups, historical societies, and even legislative tax abatement laws to encourage rehabilitation. Well organized "Barn Again" and barn preservation groups actively pursue the singular objective of saving and rebuilding as many barns as possible. Beautiful books, such as this one by Don Scott, are purchased as quickly as they are printed.

It may not be too late, after all, to save the most significant landmark ever to appear on the American scene. What a sad thing it would be if future generations never saw a stately old red barn, or ever had the pleasure of enjoying a barn's musty smells, and strange creaks and groans on a cold winter's day.

To commune with distant generations of farmers through the things that they built is the right of every American. We can be the ones who preserve that marvelous heritage for them.

preface

The stately barns that dotted the Hoosier landscape from the mid-1800s to the mid-1900s are rapidly disappearing. Changes in agricultural practices and methods, especially during the last half of the twentieth century, have diminished the usefulness of these old barns to today's farmer. As a result, many barns become an economic burden for the farmer to maintain and are left to succumb to the rigors of disuse, disrepair, rot, wind, fire, or vandalism. Others are destroyed as farms are sold to make way for larger farms, or as farms are engulfed by housing developments, urbanization, and industrialization.

Yesterday's farms were general-purpose operations that included growing assorted grains and forages and raising one or more types of livestock. Until the 1930s, much of the farming was done by horses or mules. The animals had to be sheltered and their feed protected from the environment, and barns built during the mid-1800s to the mid-1900s were constructed for that multipurpose use. The storage of animal feeds, especially hay, often required a voluminous space. Thus, most barns of that era were large structures of two or more stories, the upper being the largest and commonly called the hayloft or haymow. Today's farms are larger and more specialized. The huge agricultural implements required today do not readily fit into these barns without extensive, often cost-prohibitive, remodeling. Cattle, swine, and poultry production are more specialized and require individualized structures that are vastly different from the barns of yesteryear.

I have always had a warm spot in my heart for the farm, the soil, the animals, and the barns. Memories of the indescribable bouquet of a barn's intermingled smells of cattle, feed, and hay are still vivid. I still remember laying on hay in the hayloft, listening to the soft pitter-patter of rain on the tin roof, and dreaming of things both possible and impossible. Those days were long ago, but the memories still give me a feeling of calm.

To me, barns are somewhat like people; some have character; others are characters. Some have been taken care of and are in good condition in spite of their age; others have been neglected and have succumbed to the rigors of age and neglect. Some are elegant; others are plain. Some are workers; others are ornaments. Some are built on solid foundations; others on shaky ground. Some have great dignity in spite of age and decay. Some could tell elegant stories; others would only whisper rumors.

. . . The marks of hewing axe and adze,
Swung straight and true.
Read there the tale
Of toil and sweat and a fine pride
In shaping these great timbers.

Stand with me
A wondrous moment.
In that crafted tree
Is history enough of old great-grandsire times
A century ago and more. . . .
From "My Barn" by Dean Hughes

The whistle of the chilled wind
through the hayloft,
The soft strands of straw under you,
The pitter-patter of rain on the roof,
Creaking as the building
sways in the breeze.
An old barn is like part of history,
the memories of things within.
But, slowly it takes its place in the
ground as the soft wood rots away.
With each season
its end grows nearer,
And with each year,
Its past creeps further away.

Jesse Talley
Eighth grade student

In a November 22, 1981 *Indianapolis Star* article, Rex Redifer wrote:

American barns truly are a wonder of the world, evolving from the land and mind of man and blending with the earth so well their passing's hardly marked. As much as cathedrals and churches ever were, some barns have withstood the wind and rain of centuries; yet, no architect ever designed or built a barn. A farmer did.

Some of the interesting barn construction guidelines that have been handed down from our ancestors:

If you'd have your flatboards lay, hewe them out in March or Maye.

When the moon is new to full, timber fibers warp and pull.

Slope your roof against the northern blast; the heat of day is made to last.

It is not so much in nostalgic sense that one appreciates a barn, but with a reverence for the history of our land and the forgotten ways and skills of ancestors who produced such lasting monuments.

What is so stately as a great white barn upon a hill? So warm as that bright red barn along the road? Or so sad as the dusk-gray sagging barn beside the woods? They seem so much a part of us. And yet, they have all seen their time and soon will go the way of sailing ships.

Thus, my desire to take my hobby of photography and use it, in some small way, to preserve a heritage of days gone by, a simple document of the past for those who gaze upon these structures, as I do, for the beauty, the character, and the charm they still hold.

Donald H. Scott
West Lafayette, Indiana
January 1997

acknowledgments

My thanks to all who permitted me to walk upon their land, photograph their barns, and who cheerfully provided information on the history and colorful legends associated with their barns. Each was warm and friendly and a pleasure for me to meet. We are all indebted to those individuals who maintain their barns even though the barns may no longer be of use to them. These are, indeed, very special people. I would have liked to talk with every barn owner, but many could not be located. I am, also, indebted to the many individuals who provided information on the locations of barns that I would have otherwise not found.

My apologies to all the barn owners whose barns, for one reason or another, were not selected for use in this book. The single most important reason was not the barn or the barn owner, but rather my poor photographic technique.

To each of the following individuals I extend thanks for their help and support: Marilyn Coppernoll; Maurice L. Williamson; eighth grade teacher Jodie Clark and eighth grade student Jesse Talley at Kankakee Valley Middle School in Wheatfield, Indiana; art teacher Maryellen Cox, second grade teacher Pam Scott Hutton, and sixth grade student Chelsea Hutton at Eagle Elementary in Brownsburg, Indiana; Patti Scott Wolff, deaf education interpreter, Klondike Middle School in West Lafayette, Indiana; LaVerne Meyers, president, Jasper County Historical Society in Rensselaer, Indiana; William McVay, retired agriculture teacher, South Whitley High School in South Whitley, Indiana; the Purdue University County Extension Educators; Ron Osburn, Francesville, Indiana; and David Riggs, West Lafayette, Indiana. Last, but certainly not least, my love and my thanks to Jackie, my wife, for putting up with me for forty years. Her understanding, her inspiration, her encouragement, and her love has meant the world to me.

Right: This map of the counties and regions of Indiana is based on the crop reporting districts established by the United States Department of Agiculture. (Courtesy of the Indiana Agriculture Statistics Service Department)

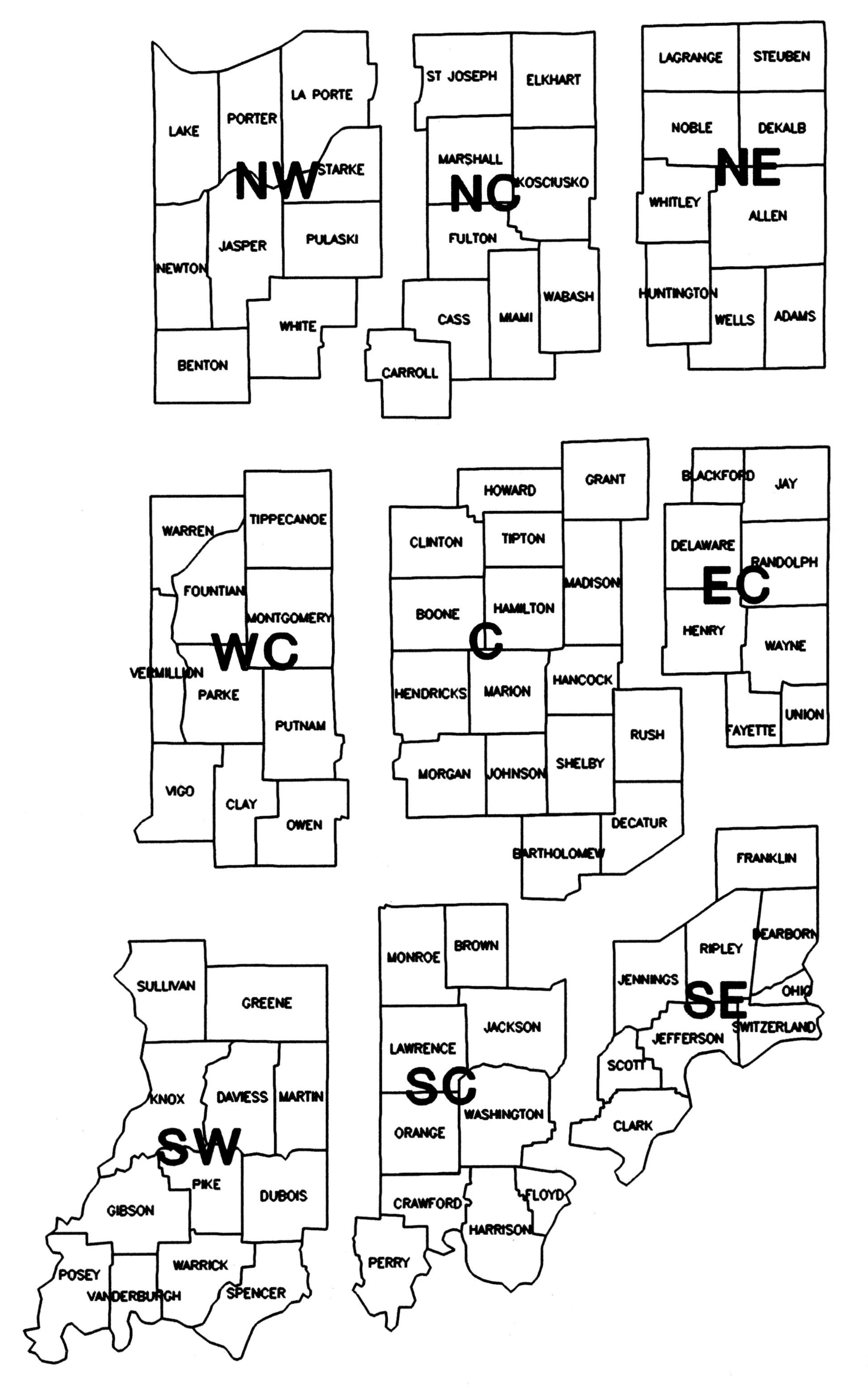
NW
LAKE
PORTER
LA PORTE
STARKE
NEWTON
JASPER
PULASKI
WHITE
BENTON
NC
ST JOSEPH
ELKHART
MARSHALL
KOSCIUSKO
FULTON
CASS
MIAMI
WABASH
CARROLL
NE
LAGRANGE
STEUBEN
NOBLE
DEKALB
WHITLEY
ALLEN
HUNTINGTON
WELLS
ADAMS
WC
WARREN
TIPPECANOE
FOUNTIAN
MONTGOMERY
VERMILLION
PARKE
PUTNAM
VIGO
CLAY
OWEN
C
HOWARD
GRANT
CLINTON
TIPTON
MADISON
BOONE
HAMILTON
HENDRICKS
MARION
HANCOCK
MORGAN
JOHNSON
SHELBY
RUSH
DECATUR
BARTHOLOMEW
EC
BLACKFORD
JAY
DELAWARE
RANDOLPH
HENRY
WAYNE
FAYETTE
UNION
SW
SULLIVAN
GREENE
KNOX
DAVIESS
MARTIN
GIBSON
PIKE
DUBOIS
POSEY
VANDERBURGH
WARRICK
SPENCER
SC
MONROE
BROWN
JACKSON
LAWRENCE
WASHINGTON
ORANGE
CRAWFORD
FLOYD
HARRISON
PERRY
SE
FRANKLIN
RIPLEY
DEARBORN
JENNINGS
OHIO
SWITZERLAND
JEFFERSON
SCOTT
CLARK

introduction

This book is simply a record of some of the barns that were still dotting the Indiana landscape during the last decades of the twentieth century. It is not one of engineering, dimensions, construction, or classification of the barns featured. It concentrates on barns built primarily between the mid-1800s and the mid-1940s.

In this book, the counties in Indiana are grouped into the same nine regions of the state that are used by the Indiana Department of Agricultural Statistics Service for their statistical reporting. The regions are designated Northeast, East Central, Southeast, South Central, Central, North Central, Northwest, West Central, and Southwest.

The photographs appearing in this book may not represent the most unique, the best maintained, the most attractive, or the most typical of the barns in a particular county or region. They were selected through the eye of the author. Also, this book is not one on photography, although it contains photographs taken primarily by the author who is still an amateur with the camera.

The classification of a barn type is often complex, especially when a barn has been modified over the years. Serious students of barn construction and history may disagree, and rightfully so, with the classification used herein by the author, which is a simplified form of that used in Noble and Cleek's *Old Barn Book*, 1995.

Please come with me, through the pages of this book, on a journey across Indiana to view some of the barns that silently stand in tribute to the ingenuity, skill, determination, and dedication of the past and present Hoosier farmers. Our journey will takes us through all ninety-two counties of the state. It starts in Steuben County, the most northeastern county in the state and winds back and forth through the counties in eastern Indiana to the Ohio River. Then it will turn northward and proceed through the central counties to the Michigan border, where it will turn southward, again, and wind its way through the western Indiana counties. We will end in Posey County, the most southwestern county in the state. You will see both similarities and differences in the barns of the regions.

Chelsea Hutton
1996

This three-gabled barn, actively used in a Steuben County dairy operation, has a full pent roof on gable end and side. Wood-louvered vents suggest the barn was built prior to 1900, but remodeled with round roof. It is located on the east side of the Steuben/Lagrange County Line Road about 0.3 mile south of County Road 475S.

Barns in Northeast Indiana

We start our journey in the northeastern part of the state where many dairy farms are still in operation. The area also contains numerous natural lakes and is known as Indiana's Lake Region. Dairy farmers frequently modify their barns that were built in the 1850 to 1950 era and continue to use them in their current operations. A number of "Plain People" communities (Amish and similar religious groups), exist in the region. "Plain People" farmers continue to work their farms in a manner reminiscent of early Indiana farmers. Horses or mules provide most, if not all, power to prepare the soil, and plant and harvest their crops. These farmers usually have relatively small, multiple-purpose farms; cattle, horses, a few chickens, and various crops such as corn, wheat, oats, hay, and pastures. They often have beautiful, well-manicured vegetable and flower gardens. Their farm buildings are usually painted white and neatly maintained. With their agricultural methods, the barns of a hundred years ago are still as useful today to the "Plain People" as they were when they were built. As a result, Northeast Indiana has many well-kept barns.

This Midwest dairy barn has a Dutch gambrel roof, silos, and milk parlor. It is a working barn in an active dairy operation. The barn, estimated to have been built in the 1930s, is located on the east side of County Road 200E about 0.1 mile south of County Road 50N in Lagrange County.

Below: This Lagrange County barn has a modified side-entry and a gable roof. The ear-corncrib and shed additions were added on gable ends later. It is estimated that the original barn was built around 1900. The barn section on the right was added much later. This working barn in an active dairy operation is located on the north side of County Road 50N at County Road 350W.

Above: This is a modified bank barn with an inclined fieldstone entry to the second floor on the side. The barn has a gable roof and a fieldstone foundation. The original barn was constructed about 1860 with hand-hewn beams and posts with pegged, mortise and tenon joints. The straw shed on left was added in 1931, making the barn into a three-gable barn. The barn is partially used in general farm operations, but maintained primarily for the benefit of the bed and breakfast operated by the owners. The barn is located on the south side of U.S. 20 west of County Road 400W in Lagrange County.

This is a large standard type barn with gambrel roof and off-center side wagon door. It is over one hundred feet in length, and currently used only for storage. The age of the barn is unknown. It is located on the south side of U.S. 6 about 0.7 mile west of State Road 5 in Noble County.

This modified bank barn has a Dutch gambrel roof and full pent roof on the side. The date of original construction is unknown, but the metal roof ventilator suggests the gambrel roof may have been a remodeling feature added after 1920. A silo is on the right and a broken windmill on the left. The barn, used for storage, is located near the junction of U.S. 6 and County Road 700N in Noble County.

These twin barns are joined together by a two-story high connection. They may have been built at different times and then connected. The construction suggests an English connected barn. The date of construction is unknown, and the foundations have been replaced. The barns are located on the west side of State Road 327 at County Road 20N in De Kalb County.

Above: This De Kalb County bank barn has a round roof and fieldstone foundation. The wagon door is off-center on the side. The concrete stave silo has a deteriorated wood shingle roof. The barn, estimated to have been constructed in the late 1800s to early 1900s, is located on the east side of County Road 9 and 0.3 mile south of County Road 16.

Left: This beautiful, unclassified barn has a gable roof, centered wagon door in the side, and two ornate cupolas. The cupolas suggest Victorian influence. The estimated construction date is about the mid-1800s. This ornate and well-maintained barn is located on the north side of State Road 37 northeast of Ft. Wayne in Allen County.

This drive-through corncrib/granary has a Dutch gambrel roof and standard cupola. The barn, estimated to have been constructed about 1930, has been well maintained over the years. It is located on the north side of State Road 37 northeast of Ft. Wayne, Allen County.

This Midwest dairy barn has silos on the left and a windmill on the right. The barn has a Dutch gambrel roof with three shed-dormers on the side, and a large hooded hay door on the gable end to the right. Metal ventilators on the roof ridge appear to be recent additions. Reportedly, this Sunnyside Farms barn was the jewel of Whitley County when built in the 1930s. This working barn is located on the north side of State Road 14 just east of Collamer in Whitley County.

This bank barn has wood-louvered ventilators, a fieldstone foundation, and centered wagon door. The barn extension on the right gives the appearance of a three-gable barn, but may have been added several years later. The barn, estimated to have been built in the mid-1800s, is located on the east side of River Road 1.7 miles south of Collamer in Whitley County.

Top Right: This three-gabled barn has a fieldstone foundation. The wagon door is on the side to the right (not shown). The barn, estimated to have been built in the late 1800s, is located on the west side of County Road 950W about 0.3 mile north of County Road 700S in Whitley County.

Right: This large modified bank barn has a gable roof, elongated wood-louvered ventilators, and remnants of large doors from the second-story threshing floor. The wagon doors to the threshing floor are located in the center of the side (not shown). The barn, estimated to have been built in the mid-1800s, is located on the north side of County Road 1100N and west of County Road 300W, Huntington County.

Below: This standard corncrib/granary has a touch of beauty in the foreground added by the farmer's wife. All buildings on this farm have been immaculately maintained over the years. The barn, constructed around 1920, is located on the east side of County Road 300W and north of County Road 1100N in Huntington County. Photo by Jacquelyn Scott.

This relatively small barn, which might be classed as a meadow barn, has a gambrel roof with wagon doors centered on the side and a pent roof on the gable end. The estimated date of construction is about 1920. The shed on right was added later. The barn is located on the west side of County Road 100W and south of County Road 100N in Wells County.

This "Plain People" type, modified bank barn has a full forbay, gable roof, and wagon doors located in the center of the side opposite the forbay. Note a lack of lightning rods on the roof ridge. Some "Plain People" groups do not believe in the use of these. This barn is located on the east side of County Road 000 about 0.5 mile south of State Road 218, Adams County.

This gable-roofed bank barn in Adams County lacks lightning rods on the roof ridge. Wagon doors, centered in the side, are protected by a porch which appears to be a relatively recent addition or replacement. The wood-louvered vents in the sides and gables suggest it was built in the late 1800s. The barn is located on the south side of State Road 218 and 0.25 mile east of County Road 000.

This bank barn has a salt-box roof, cut-stone foundation, and wood-louvered side and gable vents. The wagon door is on the side and centered. The barn, built in the late 1800s, is located on the west side of County Road 500W about 0.6 mile north of County Road 600S in Adams County.

This is a white, "Plain People," modified bank barn with full forbay, gable roof, and no lightning rods. The incline to the threshing floor is on the left side (not shown). The barn is located on the south side of County Road 300S and east of County Road 600W in Adams County.

The original portion, built in 1867, was a bank barn with a gable roof and fieldstone foundation. The straw shed on the left was added in 1904, and now gives the appearance of a three-gable barn. The incline ramp on the right leads to a center positioned wagon door on the side. The shed on the gable end of the barn was added much later. The Huntington County barn continues to be actively used in the farming operations of Mr. and Mrs. W. E. Craig, and is kept in immaculate condition.

The Craig Barn

The history of this barn, as related by Mr. W. E. Craig, goes that the original farmer, Mr. Bonebrake, and his two sons started construction of this barn in 1860, but it was not finished until after the Civil War ended. Mr. Bonebrake and his two sons were out in their woods clearing the land for fields and, at the same time, cutting the tall, straight trees already marked for the posts and beams of the barn. A rider on horseback came to them with the news that the Union Army was seeking volunteers to fight the Confederate Army. The two sons reportedly dropped their axes and immediately rode on horseback to Huntington where they caught a train to Indianapolis. They volunteered in the Union Army in Indianapolis. As the story goes, the sons knew they would only see the local area if they stayed on the farm, and they wanted to see other parts of the country. Volunteering for the Army was a way for them to see other parts of the country at little or no expense. Both sons returned to their father's farm after the War and it was then the barn was completed. The exact year the barn was completed is thought to be about 1867. Mr. Craig's father purchased the farm from Mr. Bonebrake in 1897. The straw shed was added by Mr. Craig's father.

Interestingly, all hand-hewn beams and main posts in the original barn are fourteen-inch square solid black walnut. The major roof beams are also black walnut. They are slightly smaller, but are over forty feet in length, with each being made from a single tree. In the 1850s, this farm and the general area was covered by large virgin black walnut forests, a ready source of timber for barns and houses. Many of the trees in the forest were cut, piled and burned to clear the land.

In all probability, Mr. Bonebreak had marked the trees for his barn and buildings in 1858 or 1859. It was customary for settlers of virgin lands to harvest five years of crops before they put forth the effort to build permanent structures. This conservative approach helped insure that the land would produce sufficiently to support his family. The early settlers had very little experience with the productivity of virgin soils. During this waiting period, the farmers would select large, straight trees of suitable species for the construction of their buildings. These trees were girdled and left standing in the forest for several months to two years after they died before the timber was harvested. Timber harvested in such manner was less subject to rot and insects due to the natural drying of the wood that retained the tannins and aromatic compounds within the timber.

The barn and farmstead are located at 10222N 400W, Huntington County.

Detail of pegged, mortise and tenon joints in hand-hewn timbers.

This is the interior of the Craig Barn showing the hayloft ladder. Note the hand-hewn beams and posts and that the roof rafters are small, straight, de-barked tree trunks.

Left: Detail of post and beam supports in lower level of the Craig Barn.

Original poplar siding and square nails used for attaching siding to the Craig Barn. Each nail was handmade by a blacksmith.

Close-up of pegs in a mortise and tenon joint.

Right:This was the original house built in 1861 on the site of the Craig Barn. The house has been remodeled many times over the years and served as a corncrib or granary. The original entrance door is still in the middle of the side.

Barn located at 2273E 700N, Clinton County.

Barn located at 2064N 600W Benton County.

Cupolas, Roof Vents, and Weather Vanes

Cupola on barn located on east side of U.S. 40 at Clay-Putnam County line, Clay County.

Barn located on the west side of State Road 203 north of Lexington, Scott County.

Montgomery County

Cupola on Huntington County barn.

Weather vane on barn on west side of State Road 227 about 0.6 mile north of Randolph-Wayne County line, Randolph County.

Tippecanoe County

Located on north side of U.S. 24 at County Road 1025E, Cass County.

Above: Barn located on the east side of County Road 1425W about 0.5 mile north of State Road 14, Pulaski County.

Right: Warren County barn located on the westerly side of Akers Road about 0.3 mile north of County Road 200E.

This unclassified barn has a Dutch gambrel roof, two separate wagon doors on the side, and a pent roof protecting the gable end. The shed addition on back was added later. The small building in front suggests it was the milk parlor, and the barn was used in a dairy operation at one time. The barn, estimated to have been built in the 1930s, is located on the west side of County Road 1000W and 0.8 mile north of County Road 200N in Jay County.

Barns in East Central Indiana

East Central Indiana is generally somewhat level to rolling with farming operations that raise cattle or hogs and grow crops on the more suitable land. There are many barns in this region that were built in the 1850 to 1950 era.

This small round barn is located on the east side of Indiana 1 about 0.2 mile south of Division Road in Jay County. This barn is also documented by John T. Hanou in A Round Indiana, *1993.*

This bank barn in Blackford County has a Dutch gambrel roof, wagon doors off-center on the side, and a small Dutch door to the right of the wagon doors. The estimated date the barn was built is the early 1900s. The barn is located on the east side of County Road 700E about 0.3 mile north of Indiana 26.

This is a standard barn with a gable roof and wagon door on the side and off-center. One wood-louvered vent remains with visible signs of others being replaced with metal sheeting. The vent construction and location on gable end suggest the barn was built prior to 1900. The barn is located on the west side of County Road 700E about 0.3 mile south of County Road 200N in Blackford County.

This is a modified standard barn with a broken-gable roof, cut-stone foundation, and wood-louvered vents on both gable ends and sides. The wagon door is centered on the side. The barn, estimated to have been built in the late 1800s, is located on the east side of County Road 1050E about 0.6 mile north of Indiana 28 in Delaware County.

Left: This is a modified bank barn built in 1867 with gable roof, extended roof ridge for hay hood, plank and batten siding, and centered side wagon doors. The original cut-stone foundation has been replaced. The barn is still actively used in the farm operation. All buildings and fences on this farm have been magnificently maintained. The barn is located on the north side of County Road 700N about 0.2 mile west of Indiana 67 and 28 in Delaware County.

Below: This is a standard barn with a Dutch gambrel roof, cut-stone foundation, hooded hay door, and concrete stave silo. The wagon door is centered on the side (not shown). The shed on the right was added later. Note the hay door is not hinged at the bottom. Rather, it is counterweighted and slides up and down like a double-hung window. The barn is located on the north side of Indiana 28 about 0.6 mile east of County Road 100W in Delaware County.

This round roof barn was built in 1953 after the original barn was destroyed by a tornado. The laminated roof rafters were custom made. The wagon door, hay door, and hay hood are on the gable end. There is, reportedly, only one other barn in the United States like this barn, and it is in Missouri. The barn is located on the north side of State Road 67 and 28 about 0.2 mile east of County Road 700N in Delaware County.

This standard barn in Randolph County has a Dutch gambrel roof and wagon doors centered on the side. It has a gabled roof dormer and a circular vent/window above the wagon doors. The date the barn was built is unknown. The barn is located on the south side of County Road 500S about 0.5 mile west of State Road 227.

Left: This neat bank barn has a gable roof, fieldstone foundation, and post-supported bays on the left. The wagon doors are centered on the side. The original rectangular vents and light source above the wagon doors have been replaced with windows. The barn, built in the late 1800s, is located on the west side of State Road 227 in Randolph County about 0.6 mile north of the Wayne County Line.

Below: This barn is one of the older structures on the Davis-Purdue Agricultural Center farm in Randolph County. It is a standard barn with a gable roof and wagon door on gable ends. The shed extensions on the sides were added later. The barn has been modified several times over the years. The Davis-Purdue farm is on the west side of State Road 1 about 1.5 miles south of State Road 28.

This older bank barn has a gable roof, wooden cupola, and fieldstone foundation. The wagon doors are centered on the side. Some of the original wood-louvered vents remain on the gable end and on the cupola. The barn, estimated to have been built around the 1860s to 1870s, is located on the south side of State Road 38 at Manning Road in Wayne County.

This is a bank barn with a gable roof and wagon doors off-center on the side. Post-supported bays are in the gable end. A few wood-louvered vents remain on the gable end. Note the scalloped, vertical siding beneath the eaves. The original foundation has been replaced. The barn, built in the late 1800s, is located on the north side of Beelor Road and just west of U.S. 27, Wayne County.

This standard barn has a wooden cupola, remnants of wood-louvered vents, and a gable roof. Date of original construction is unknown. The barn is located on the west side of County Road 600E about 0.3 mile south of State Road 36 in Henry County.

This is a standard barn with paint similar in color to the original "turkey red" color. It has a gambrel roof, and the wagon door is centered on the side. It is not known if the extension on the right was added at time of original construction. The hayloft door was originally a counter-weighted door that moved up and down like a double-hung window. The estimated date of construction is the late 1800s. The barn is located on the north side of U.S. 36 about 0.3 mile east of State Road 3 in Henry County.

This scenic, modern horse barn is on a farm ringed with new white fencing. The barn is located on the west side of County Road 550E about 0.1 mile south of Base Line Road in Fayette County.

This older barn, with a monitor roof and the remnants of wood-louvered vents, is no longer used. Its original fieldstone foundation has been partially replaced. The date the barn was built is unknown, but the wood-louvered vents suggest it was prior to 1900. The barn is on the north side of Old Brownsville Road just east of Brownsville in Union County.

This round barn is located on the south side of County Road 50S just west of Round Barn Road in Union County. It is further documented by John T. Hanou in A Round Indiana, *1993.*

This modified bank barn has a gable roof and enclosed porch protecting the gable-end wagon doors. The barn, estimated to have been built in the early 1900s, is located on the south side of Golden Road about 0.1 mile east of State Road 101 in Franklin County.

Barns in Southeast Indiana

Many areas of the Southeast Region are rolling to hilly. Some are heavily wooded. Cattle and hog production are significant agricultural enterprises. Crop production occurs in areas suitable and tobacco production occurs on upland soils. Many barns are unpainted and generally tend to be somewhat smaller than their northern counterparts.

Right: This is a series of cattle barns. The nearest is a modified bank barn with a gable roof and unusual metal vent on the roof ridge, estimated to have been built in the 1920s. The barns are located on the northeast corner of Holland Road and U.S. 52 in Franklin County.

This standard barn, built in 1901, has a gable roof, wooden cupola, and cut-stone foundation. The wagon doors are centered on the side. Originally it was a cattle barn, but currently is used for storage of small equipment and wood. The barn is situated on the south side of State Road 48 just west of North Hogan Road in Dearborn County.

This is a standard barn with a gambrel roof and wagon doors centered on the side. The cut-stone milk parlor was added later, and the aluminum siding was recently applied. Note the metal owl decoration just beneath the widow on the gable end. The barn, no longer used as a dairy barn, is estimated to have been built about 1920. It is located on the east side of Weissburg Road about 0.4 mile north of State Road 48 in Dearborn County.

This modified standard barn in Ripley County has a gable roof and wooden cupola. Note the star decoration and the wood-louvered vent above the wagon doors centered on the side. The wood-louvered vent suggests the barn was originally built in the mid- to late-1800s. The cupola appears to have been replaced recently. The concrete stave silo and connecting shed were added later. The barn is located on the west side of State Road 101 about 0.2 mile north of County Road 1200N.

This small standard barn with its rusting gable roof and metal silos has a certain charm. The barn is no longer actively used, and its date of construction is unknown. It is located on the west side of State Road 56 north of Rising Sun in Ohio County.

Left: This relatively large bank barn with a gable roof, wooden cupola, and cut-stone foundation is now abandoned. It is located on the north side of State Road 156 at the east edge of Vevay in Switzerland County.

This three-gabled barn originally had an unusual, nonmortared cut-stone foundation, although much of it has been replaced. The hay door and enclosed hay hood are on the gable end. The 16-pane windows are unusual in Indiana barns. The estimated date of construction is about the mid-1800s. The barn is located at 3355N State Road 7 in Jefferson County.

Right: Detail of the nonmortared cut-stone foundation.

The vertical planks immediately above the wagon doors were added some time later. These planks replaced the original wood-louvered vent and source of light which was used in many of the older barns. This abandoned barn is located on the east side of State Road 7 about 0.3 mile south of County Road 500N in Jefferson County.

This large working cattle barn with a monitor roof is situated on the north side of U.S. 50 at the west edge of North Vernon in Jennings County. The date of construction is unknown.

Right: This huge building is a modified bank barn with hip roofs and multiple dormers. The barn is on the Englishton Park Retirement Community Childrens Retreat located on the west side of State Road 203 about 0.2 mile north of Lexington in Scott County.

Left: This small meadow barn has a gambrel roof, with a hay hood and wagon doors on the gable end. No additional information was obtained on the barn which is located on the north side of State Road 356 about 1 mile east of State Road 3 in Scott County.

Right: This large, Midwest dairy barn was built in 1947; possibly one of the last barns of this style to have been built. It is of concrete block construction with laminated roof rafters and a Dutch gambrel roof. The barn is no longer used in a dairy operation. The metal stairs on the left lead to the hayloft that now houses a full-sized basketball court complete with lights, water fountains and other accouterments. The barn is located on the east side of Oscar Long Road about 0.2 mile north of State Road 62 in Clark County.

This Clark County barn outlived its usefulness as a farm barn, but was extensively remodeled and converted to a restaurant and gift shop for a highly successful family orchard, winery, and tourist business near Starlight.

This is a true tobacco barn with gable roof and wagon doors on both gable ends. Note the several round roof ventilators and the hinged, vertical planks on the side that can be opened to increase air circulation to cure the tobacco. This barn was built in the 1940s. The barn is located on the west side of King Road about 0.4 mile north of Old Salem Road in Clark County.

Right: This interior photo of the tobacco barn shows the method of hanging the tobacco for drying.

This is a drive-through crib/granary with gable roof. The sills of the crib sit atop cut-fieldstone piers. The crib for ear corn is on the right side with the grain bins on the left. This crib was built in the late 1800s and is on the same farm as the tobacco barn.

This Clark County log cabin was built in 1825, and has been wonderfully maintained by the descendants of the original builder as a remembrance of days gone by. It was the first home for the Hilton family that purchased the land in 1825. The farm, the same one with the tobacco barn, has been in the family since that time.

Left: While not a barn, this type of structure played an important role in the early days on Hoosier farms. It is one of the few remaining smokehouses left in the state. Farmers hung their butchered meat in the smokehouse and then cured it with a combination of smoke, salt, and other ingredients. This well-maintained structure, on the same farmstead as the tobacco barn, was built in the 1800s.

This very rare "Old Loyalty" advertisement is on a Ripley County barn located on the east side of State Road 421 about 0.5 mile north of County Road 450S.

Right: This Switzerland County barn has a rare "Chew Mail Pouch" advertisement on the side and a "Smoke Kentucky Club" advertisement on the gable end. The barn is located on the north side of State Road 156 west of Lamb.

Left: "Jesse James Hideout" is advertised on this old Crawford County barn on the south side of State Road 62 west of Corydon.

Barn Advertisements

Below: This Harrison County barn is the last "Chew Mail Pouch" advertisement painted by the famous Harley E. Warrick, painter of nearly all "Chew Mail Pouch" barns in the country. It is located on the east edge of Lanesville, Harrison County.

Right: One is advised to "Chew Mail Pouch Tobacco" on this Cass County barn located on the north side of U.S. 24 about 0.2 mile east of County Road 450W.

Below: The advertisement on this old bank barn is the name of the business that now occupies the building in Corydon. The barn, estimated to have been built in the mid-1800s, has hand-hewn timbers, pegged mortise and tenon joints, gable roof, wood cupola, and cut-stone foundation. The Harrison County barn has been extensively modified to house the antique business.

This barn on the north side of U.S. 6 at the east edge of Kendallville, Noble County has one of the few remaining "Meramec Caverns" advertisements.

Left: Advertisements cover most of this Jefferson County barn located on the southeast corner of State Roads 62 and 250.

This Floyd County raised barn sits on the north side of State Road 62 about 0.5 mile west of Kensington Road. No additional information was obtained about this barn.

Top Right: This bank barn was moved to this location and refurbished by the Lanesville Historic Society. It has a gable roof, wood cupola with wood-louvered vents, and a wagon door centered on the side. The barn, which had a cut-stone foundation, is estimated to have been originally built in the 1850s to 1860s. Note the gable-roofed dormer above the wagon door. The barn is located on the north side of State Road 62 at the east edge of Lanesville in Harrison County.

Barns in South Central Indiana

South Central Indiana is characterized by a rolling to hilly topography and heavily wooded in many areas. True to tradition of the early settlers, many barns in this region have never had a drop of paint. While there are some interesting barns in the region, they are difficult to find because of the wooded, hilly terrain. Cattle production occurs on the rolling areas, with crop production in the more level river bottoms and valleys. Tobacco production occurs on the upland soils.

This older barn has many wood-louvered vents on the side. The barn has been modified, especially the gable end on the right. The vents, with original fieldstone foundation, and the beams and posts suggest the barn was built in the late 1800s. The barn is located on the north side of Bradford Road about 0.1 mile east of State Road 135 in Harrison County.

This is a large standard barn with a gable roof, fieldstone foundation, and hooded hay door. Wagon doors are on the side and off-center. This barn, constructed on an unknown date, is no longer actively used. The barn is located on the south side of State Road 62 and east of State Road 66 in Crawford County.

This huge, modified bank barn in Perry County has a gable roof and posted side pent roof. The wagon door and drive incline are on gable end to the left (not shown). The cut-stone foundation is made of large sandstone blocks. This barn was covered with composition shingles several years ago. The foundation, the beams and posts, and the remnants of wood-louvered vents suggest that the barn was built in the mid- to late-1800s. The barn, which is not in use at this time, is located on the north side of State Road 66 about 0.5 mile east of the Perry-Spencer County line.

Above: This large, standard barn has a Dutch gambrel roof, cut-stone foundation, hay door, and hay hood. The wagon door is on the gable end (not shown). The barn, built on an unknown date, is located on the east side of State Road 66 about two miles south of Derby in Perry County.

Right: This is a modified bank barn with a gable roof, cut-stone pier foundation, and wood-louvered vents on sides and gable ends. The wood-louvered vents suggest the barn was built in the late 1800s. The barn is located on the east side of State Road 337 at the south edge of Bromer in Orange County.

Left: This large standard barn in Orange County has a gable roof and numerous wood-louvered vents. The vertical siding on the gable end suggests this was originally a three-story barn. The hay door is on the gable end, and the wagon door is off-center on the side. The barn, estimated to have been constructed in the mid- to late-1800s, is located on the east side of State Road 337 at County Road 625N.

This small, extended barn has a broken gable roof, remnants of wood-louvered vents, and a fieldstone foundation. The wagon door is in the gable end. The barn, estimated to have been constructed in the late 1800s to early 1900s, is located on the east side of State Road 60 about three miles north of Pekin in Washington County.

This large cattle barn has three bays, a monitor roof, and an extended gable peak protecting the hay door. Part of the foundation has been replaced with poured concrete pillars. The age of the barn is unknown. The barn is located on the north side of State Road 160, east of Salem in Washington County.

This extended barn, of more recent vintage, has a broken gable roof, and hooded hay door. The working dairy barn is located on the east side of State Road 37 at Guthrie Road in Lawrence County.

This is a relatively small standard barn with a gambrel roof, fieldstone foundation, and a nonprotected hay door. Shed extensions on either side of the barn appear to have been added later. It is estimated the original barn was built in the late 1800s in Jackson County. The barn is located on the north side of County Road 1100N about 0.1 mile west of State Road 135.

This is a relatively large extended cattle barn with a monitor roof and hooded hay door. The wagon doors are on the gable ends. It is estimated to have been built in the early 1900s. The barn is located on the north side of Wampler Road at County Road 800W in Monroe County.

Right: This relatively small standard barn was built in 1880. The concrete silo and powerhouse were constructed later. The barn has a typical gable roof, hooded hay door, and gable-end wagon door. The barn is located at 8102N Bottom Road in Monroe County.

This feeder barn is nestled in a rich creek-bottom valley in Brown County. The barn has a typical gable roof with low eaves. The sheds on either side of the barn appear to have been added later. The black color may have been original or added several years ago. The barn, estimated to have been built in the late 1800s, is located on the westerly side of State Road 135 about two miles south of Story.

This relatively small standard barn has a gable roof, hooded hay door, and a gable-end wagon door. The barn siding is of vertical plank and batten construction. The original fieldstone foundation has been replaced with concrete blocks. The barn, estimated to have been built in the late 1800s, is located on the westerly side of State Road 135 at the northern edge of Story in Brown County.

This standard barn has a Dutch gambrel roof with hay and wagon doors on the gable ends. It has been modified for hog production. The barn, estimated to have been built in the 1920s, is located on the north side of Old State Road 44 west of Rushville in Rush County.

Top Right: This older Decatur County barn is east of Greensburg on the south side of Base Road about 0.3 mile west of County Road 350E. This barn had a cut-stone foundation which has been partially replaced by poured concrete. The concrete stave silo has a relatively rare conical, wooden roof and enclosed silo access. It is a standard barn with a broken gable roof and gable-end wagon door. It is not known if the broken gable roof was the result of shed additions on the right and left, or if these were part of the original barn. The roof vents are later additions and somewhat unusual, although other barns in the area have similar vents. The estimated date of construction is about the mid- to late-1800s.

Barns in Central Indiana

The Central Region of the state contains sixteen counties, and is the largest region in the state. The region has a widely diversified agriculture, highly diversified topography, and variable soil productivity. The northern portion of the region has some of the more productive hog- and crop-producing areas in the state. Urban and industrial development have virtually obliterated agriculture in Marion County and are impinging on the surrounding counties.

This sills of this old, standard barn sit atop cut-stone piers. The gable roof was extended for the hooded hay door and vent. The wagon door is on the opposite end (not shown). The original barn was built in Bartholomew County about the mid- to late-1800s. The shed on the left was added much later.

This working Shelby County dairy barn is located on the east side of County Road 250E about 0.4 mile south of County Road 750S. It is an extended barn with a gambrel roof and the windows above the side extensions give it the appearance of a monitor roof barn. The hay door and hay hood are in one gable end. The wagon doors are on both gable ends. The date the barn was built is estimated to be about 1920 to 1930s.

This Rush County bank barn, built in the late 1800s, has a gable roof with wagon door in the gable end. Its cut-stone foundation has been largely replaced. Aluminum siding has been placed over the original siding and the original hay door and circular vent have been covered. The well-maintained barn is located at 6153N 50W.

Right: This standard barn has a gable roof and centered, side wagon door. Some of the original wood-louvered vents remain on the sides and gable ends. The hay door is in the gable end (not shown). The original foundation has been replaced with concrete blocks. The barn, estimated to have been built in the mid- to late-1800s, is east of State Road 9 near Shelbyville in Shelby County.

This large, standard barn will soon be engulfed by a housing development or industrial business in Johnson County. It is located on the north side of County Road 500N about 0.5 mile west of County Road 525E. The Dutch gambrel roof, hay hood, shed dormers, and the off-center wagon door on the side suggest the barn was built in the late 1800s to the early 1900s.

Encroaching urbanization will soon engulf this Johnson County barn. The vertical plank and batten siding suggest this barn was built in the late 1800s. It has a gable roof, hooded hay door, and the wagon door is in the gable end. It is located on the west side of County Road 500E just north of County Road 100N.

Left: This small, attractive Morgan County barn with a gable roof, hooded hay door, and wagon door in the gable end, is currently being used for horses. This barn is estimated to have been built in the early 1900s. It has been made more attractive by the recent addition of the aluminum siding and roof. It is located on Leonard Road about 0.6 mile west of State Road 252.

Below: This working, extended cattle barn, estimated to have been built in the early 1900s, is located on the south side of County Road 400S about 0.3 mile east of U.S. 31 in Johnson County. The barn has a broken gambrel roof, hooded hay door, and wagon door on the gable end.

This working Morgan County barn, estimated to have been built in the late 1800s to early 1900s, is located on the north side of State Road 144 about 0.5 mile west of the West Fork of the White River. It is a standard barn with a wooden cupola, wood-louvered vents in the cupola, Dutch gambrel roof, hay hood, and wagon doors in gable ends. The original fieldstone foundation has been mostly replaced with poured concrete. The shed was a later addition and the silo roof is gone.

This is a relatively small standard barn with gable roof and enclosed porch on the gable-end wagon door. This barn, built about 1900 to 1910, is located on the west side of State Road 39 at County Road 500N in Hendricks County.

Left: This unique dairy barn is constructed of concrete and concrete blocks. The side extensions have gable roofs while the main portion has a gambrel roof. The barn, no longer used in a dairy operation, is estimated to have been built about 1920. It is located on the north side of Southport Road just west of State Road 37 in Marion County.

Below: This standard barn in Hendricks County has a gable roof and a partially enclosed porch for the wagon door which is centered on the side. The date the barn was built is unknown. The barn is located on the south side of State Road 136 just west of Lizton.

Right: This Marion County barn was built in 1880. The beams and posts are hand-hewn with pegged, mortise and tenon joints. The ornate circular vent in the gable peak with wood-louvered vents and original fieldstone foundation make this barn unique. The extensions on the front and sides of the barn appear to have been added later. The barn, no longer used in a farming operation, is located on the south side of U.S. 52 just west of the Marion-Hancock County line.

Left: This beautiful barn is the largest round barn in Indiana and may be one of the most photographed in the state. It has been maintained in immaculate condition and is still used in an active farm operation. The barn was built in 1902–1903, and is located on the north side of County Road 600N about 0.2 mile east of County Road 500W in Hancock County. This barn is also documented by John T. Hanou in A Round Indiana, *1993.*

Right: This beautiful, large barn in Hancock County is no longer used in a farming operation, but has been well maintained and preserved. It was built as a horse/mule barn for the animals used in the farming operation and housed over forty animals. The oval, wood-louvered vents have been replaced with windows. The metal vents on the roof are also of more recent vintage. The barn, built in the late 1800s to early 1900s, is located on the south side of U.S. 40 between Greenfield and Indianapolis.

This Madison County standard barn has been modified several times. The original barn was built during the early 1900s. The cupola and shed on the right were more recent modifications. The barn is located on the east side of State Road 13 about 1 mile south of Lapel.

This three-gabled barn is located on the north side of State Road 38 about 0.1 mile east of County Road 875W in Madison County. The date of original construction was not determined. This barn has been extensively changed over the years, and is typical of many Hoosier barns that have been modified to be more useful in modern farming operations.

This mid-1800s bank barn has hand-hewn timber framing, cut-stone foundation, wood-louvered vents, and a hay hood. The wagon door is located in the gable end. The exact date the barn was built is unknown, but is estimated to be around 1860. It is located on the north side of State Road 38 just east of Little Chicago Road in Hamilton County.

This is one of the relatively few all brick barns in the state. It was built in 1932, has a Dutch gambrel roof, and a wagon door on the side. Currently used as a cattle barn, it is located on the south side of State Road 32 at the west edge of Eagletown in Hamilton County.

Left: This large working barn is close to being a raised barn. It has a gambrel roof, three gable-roofed dormers on either side, full pent roofs on side and gable end, and a hay door in the gable end (not shown). It is estimated this barn was built around 1910 or 1920. An older barn is partially shown to the left. The barns are located on the north side of 206th Street in Hamilton County.

Right: This large feeder barn is located on the west side of State Road 39 just south of County Road 200S in Boone County. The side of the barn is constructed of concrete blocks which were made to resemble cut stone. Only one of three metal vents remain. The hooded hay door and wagon door are located in the gable end. It is estimated the barn was built around 1920 to 1930.

Left: This Boone County standard barn has had the original hay door replaced by a smaller door located lower on the gable end to accommodate a mechanical elevator for the handling of baled hay. The barn has a gable roof with the wagon door in the gable end. The barn, estimated to have been built about 1920, is located on the east side of the Boone-Mongomery County Line Road and just north of State Road 32.

Below: This photo shows four different types of farm structures with different erection dates. The small frame barn on the far right is an equipment storage building, probably built in the 1940s. The wire, corncrib just to the left was erected when corn was harvested on the ear, probably around the mid- to late-1930s. The gray granary to the left of the crib is the newest of the four structures, built probably in the 1950s or 1960s. The metal grain bin to the far left was probably built in the late 1940s to mid 1950s. These structures are located on the east side of County Road 175W just south of County Road 650S in Boone County.

This beautiful, well-maintained, pegged, hand-hewn timber beam and post barn is in Clinton County. The two-tiered cupola with its wooden spire and wood-louvered vents is unusual in the state. Note the elongated wood-louvered vents on the gable ends and the circular wood-louvered vents on the side. The upper gable end vents have been replaced with windows. The roof is of gable design and the wagon door is off-center in the side. The exact date the barn was built was not determined, but estimated to be during the mid-1800s. The barn is located on the south side of County Road 700N about 0.3 mile east of County Road 200E.

This barn found a new life after remodeling. It currently serves as home for the Red Barn Summer Theater near Frankfort. Originally, it appears to have been a standard barn with a gambrel roof and wagon doors in the gable end. The barn is located on the north side of County Road 150S about 0.4 mile west of County Road 280E in Clinton County.

The lower portion of this working Midwest dairy barn is glazed tile. The barn has a Dutch gambrel roof, shed dormers, a hay door in the gable end, a hay hood, and attached milk parlor. It also has both a concrete stave and poured concrete silos. The barn, estimated to have been built about 1930, is located on the west side of County Road 950W about 0.3 mile north of County Road 350N in Clinton County.

This round barn is unusual because it was built without windows (Hanou, 1993). It is no longer a working barn. The barn is located on the south side of County Road 25E and south of County Road 300S in Tipton County.

Right: The colors of this weathered Tipton County barn add character to a building that has very little utilitarian value to a grain farmer. The centered wagon door on the side is too small to permit entrance of today's huge agricultural equipment, and the cost of remodeling is too great. The future of this old barn with its Dutch gambrel roof and hooded hay door is in question. It is located on the west side of County Road 1175W about 0.1 mile south of County Road 450N.

Left: This standard barn with a gable roof and centered wagon door on the side is no longer used for the purpose for which it was built. As the advertisement indicates, the barn will, temporarily at least, be part of a golf course and housing development project. The barn is located on the north side of State Road 26 about 1 mile east of County Road 200W in Howard County.

This Grant County barn identifies both the name of the owner as well as the date the barn was built. This bank barn has a gambrel roof and a gable-end wagon door. Although the barn was built in 1905, it has been remodeled and recently covered with aluminum siding. The barn and silos continue active use under direction of the seventh generation of the same family. The barn is located on the north side of State Road 26 just east of Fairmount.

This somewhat unusual concrete and concrete block bank barn, thought to have been built about 1930, is located on the same farmstead as the preceeding photo. The concrete blocks were made to resemble fieldstone blocks.

This structure built around the 1850s is a three-gabled barn with the 'L' section being hidden behind the front section shown here. This barn has five cupolas; three seen here and two more on the 'L'. The large cupola on the right has an unusual roof. Many of the wood-louvered vents remain. The gable roof and extended roof ridge forms the hay hood. Note there are several hay doors. The siding is the vertical plank and batten strip construction, and the beams and posts are hand-hewn timbers fitted together with pegged, mortise and tenon joints. Quite a jewel in its day. The barn is located on the west side of County Road 1150E about 0.3 mile north of County Road 600S in Grant County.

A dinner bell is mounted on the gable of this small barn in Franklin County. Note the wood-shingle roof. Barn is located on the south side of Golden Road about 0.1 mile east of State Road 101.

Right: Window, vent, and cupola decorations enhance this Clinton County barn.

Barn Decorations

An ornamental hex sign decorates the gable on this Porter County barn located on the west side of State Road 2 about 0.3 mile south of County Road 150S.

Right: Window treatment. An older wood-louvered vent has been replaced by a window on this Tippecanoe County barn on the west side of County Road 900E about 0.2 mile south of County Road 50S.

Left: Window treatment, in Clinton County, that has retained a portion of the original wood-louvered vent. Barn is located on the south side of County Road 700N about 0.3 mile east of County Road 200E.

Right: This old circular wood-louvered vent is on the Clinton County barn above.

Flowers, an American flag, and decoratively painted doors, windows, and vents have beautified this Wabash County barn located on the east side of County Road 300E about 0.1 mile south of State Road 114.

Left: Remnants of decorative trim remain around this mid-1800s, long, wood-louvered vent on this Huntington County barn at 3736W 1100N.

Below: These Posey County grain bins are painted to resemble tea, coffee, sugar, and flour canisters. They are located on the west side of County Road 700E about 1 mile north of State Road 62.

The Breaks-Myers barn in Montgomery County.

The Breaks-Myers Barn

The barn was built by the Wilbert Breaks family. Construction was started in the summer of 1912 and finished in April 1913. The building architect was a Mr. Sharp with George W. Stout master carpenter. Mr. Stout was also master carpenter for the Whitesville School and many homes in the area. He was not known as a builder of barns and reluctantly took on this project. He was selected for his meticulous precision in all details of building homes and schools. His mastery and precision are, to this day, still evident in the interior details of the barn.

The site selected for the barn happened to be in a low-lying area and the grade was increased by three feet with cinders hauled with horses and wagons from the electric light plant in Crawfordsville. The farm was purchased from the Breaks by the Worley Myers family a few years later. The farm continues in the Myers family and the barn is still actively used in their farming operations.

The barn is a fourteen-sided structure, not a true round barn as it closely resembles. It was the jewel and landmark of the area in the early years. But, with time and weather, deterioration occurred and the barn became ragged. In 1981, the barn was extensively repaired, a new roof added, and the sides were covered with aluminum. Then, once again, the barn regained its jewel and landmark status.

The barn is also documented by John T. Hanou in *A Round Indiana*, 1993, but the original owner is listed as Wilbur Breeks rather than Wilbert Breaks.

This old log barn was razed to make way for the fourteen-sided Breaks-Myers barn. Photo courtesy of Jeff Myers.

The date of the old log barn's construction is unknown. It was built with debarked, notched logs of varying sizes fitted together by notched ends. They were not hand-hewn, and a 1958 letter from Mrs. Breaks states that many of the logs were sassafras. Photo courtesy of Jeff Myers.

This is the new Breaks-Myers barn shortly after construction in 1913. Note the cinder fill used to raise the grade visible around the outside of the barn. Photo courtesy of Jeff Myers.

Portions of an old stationary thresher in operation are visible on the left of the Breaks-Myers barn. Note the windows beneath the wood-louvered vents in the cupola and dormers were temporarily removed to improve air circulation. Photo courtesy of Jeff Myers.

Left: Renovation of the deteriorated Breaks-Myers barn was started in 1981. Photo courtesy of Jeff Myers

Below: Wagons loaded with grain wait to be threshed at the Breaks-Myers barn. Portions of an old stationary steam thresher are on the right. Photo courtesy of Jeff Myers.

This somewhat unusual two-level barn was probably used for poultry production at one time. It has a concrete foundation, gable roof, wagon door in the gable end, and twin cupolas. The barn, estimated to have been built in around 1910–1920, is located on the west side of State Road 13 about 0.1 mile south of County Road 800N in Wabash County.

Top Right: This is a beautiful, well-maintained Wabash County modified bank barn. The barn has a broken gable roof, twin wagon doors on the side, and elongated wood-louvered vents. Note the unusual serrated metal roof ridge plate. The barn, estimated to have been built in the mid- to late-1800s, is located on the east side of County Road 300E about 0.1 mile south of State Road 114.

Barns in North Central Indiana

The North Central Region has generally level to gently rolling topography with highly fertile soils in the southern portion and crop and hog production are major agricultural enterprises. The northern portion becomes somewhat more undulating and dairy production occurs.

This unusual three-gabled barn has decorative, scalloped, vertical siding just beneath the eaves. The gable doors are located on the gable end of the long section (not shown). The barn was built in 1903 with side additions apparently built later. The barn is located on the northwest corner of State Road 13 and County Road 1050N in Wabash County.

This 1860s brick barn served as a wintering barn for the big cats of the Wallace Circus. Reportedly, their claw marks are still visible on the walls. The barn is somewhat unusual in that it is all brick construction. It has a gable roof, centered side wagon door and wood-louvered vents. The section to the right was a much later addition. The barn is located east of Peru on the north side of State Road 124 about 0.9 mile west of County Road 625E in Miami County.

This huge Miami County barn was, at one time, owned by Cole Porter, famous American song writer. It is rumored that this barn was used as a wintering barn for circus elephants. The gable roof is adorned with unusual vents, but these may have been added later. Note the lattice-style wood vents just beneath the eaves and the wood-louvered vents on the gable end. It has a hooded hay door with wagon doors in the gable end as well as in the sides. This barn, estimated to have been built in the mid- to late-1800s, is located on the south side of State Road 124 about 0.5 mile west of County Road 300W.

The story goes that the name Good Enough was derived from the original farmer who brought his new bride home and was somewhat embarrassed by the shabby appearance of his Miami County farm. His bride's comment was simply, "It's good enough for me." Thus, when this barn was built in 1901, it was named the Good Enough barn. The barn has a Dutch gambrel roof, hooded hay door and wagon door on the gable ends. Although not readily apparent in this figure, the side areas just beneath the eaves are completely open. The vertical siding stops about three feet from the top beneath the eaves. This barn is still in active use in the farming operation. The farm's owners, Sidney Kubesch and his family, were awarded the John Arnold Rural Preservation Award during the 1993 Indiana State Fair for their preservation of this farmstead and its buildings.

Center: This all brick barn, on the same farm as the Good Enough barn, was built during the early 1900s. It has a Dutch gambrel roof with twin, shed-dormers, hooded hay door, and centered side wagon door. It continues as a working barn. The Kubesch farmstead sits just north of State Road 124 and near the Peru Circus Museum in Miami County.

This well-maintained granary is also on the Kubesch farm. As most early granaries, the sills sit atop pillars with metal strips to prevent rodents from entering.

Right: This standard barn with a gable roof, wooden cupola, and twin wagon doors in the side was built in the 1800s; exact date not determined. This Cass County barn is located on the south side of County Road 750S about 0.3 mile east of Cass-Miami County Line.

Below: The metal roof vents of this Cass County barn are unusual. This is a standard barn with a gable roof, concrete foundation, and off-center wagon door. The shed on the right appears to have been a later addition. The barn is located on the north side of old U.S. 24 at County Road 1025E. The road may no longer be in existence due to the construction of new U.S. 24.

This Cass County barn was built about 1880. It is a standard barn with a gable roof, hay door and hay hood in the gable end (not shown), and wagon door off-center in the side. The shed addition on the left was added in 1940. The barn is located on the north side of County Road 800S about 0.2 mile east of the Cass-Miami County Line.

This attractive, well-maintained bank barn was built on the upland just above the Wabash River flood plain. It has a gable roof, wood-louvered vents, and off-center wagon door in the side. It is presumed the original foundation was cut fieldstone. The barn was built in 1859 with hand-hewn beams and posts with pegged mortise and tenon joints. The barn sits on the south side of Bicycle Bridge Road just west of County Road 925W in Carroll County.

This Carroll County house was originally a dairy barn built in the early 1900s. It was completely renovated into a home in the 1970s. The interior beams, posts, walls, and floors throughout the structure were cleaned, scraped, and sand blasted, then finished in the natural wood. The hayloft is now the main living quarters. Quite a beautiful home; it sits on the north side of Bicycle Bridge Road almost in site of remnants of the Wabash-Erie Canal to the southeast, and it is northwest of Delphi.

While technically not a barn, this 1845 flouring mill has strong connections to the early history and agriculture of the area. The mill was restored in 1940 by Mr. and Mrs. Claude Sheets. The Wild Cat post office was in the mill from 1850 to 1894. The Wild Cat Masonic Lodge organized here in 1864 and met on the third floor until 1867. The mill is near Cutler on County Road 50E about 0.5 mile north of County Road 500S in Carroll County.

This relatively large standard barn, still in active use, has been recently clad with aluminum siding. The barn has a gable roof and wagon doors off-center on the side. The barn, estimated to have been built around 1920, is located on the north side of State Road 114 just east of County Road 300W in Fulton County.

This dairy barn has the design of a Holstein cow on the roof made from colored composition shingles. There are many barns in Fulton and Miami Counties with roof designs such as this. Photographs of this particular barn have appeared in several publications and magazine articles. It is located on the west side of County Road 200W about 0.1 mile north of County Road 900S in Fulton County.

Right: This tall, Marshall County bank barn continues use as a horse barn. Note the horseshoes painted on the gable end, the twin doors just above the full pent roof, and the unusual metal roof vents. It has a Dutch gambrel roof and the wagon door is centered in the side (not shown). The barn is located on the north side of Destiny Road at the north edge of Breman.

Left: This modified bank barn has a gable roof with the wagon door off-center in the side. Its weathered, poured-concrete foundation suggests this barn was built in the early 1900s. The metal roof ventilators may have been added later because they did not become popular until about 1920. The sheds on the left and rear of the barn may have been later additions. The barn is still in active use, and is located on the northwest corner of State Road 331 and 11th B Road in Marshall County.

This 1850s bank barn is still in active use in a Kosciusko County dairy operation. The bank and incline drive to twin, off-center wagon doors are on the side (not shown). The gable roof is supported by rafters made from straight three- to five-inch diameter trees which still have the original bark on them. The posts and beams are hand-hewn timbers with pegged, mortise and tenon joints. The original fieldstone foundation has been largely replaced. The barn is located on the east side of State Road 13 about 0.1 mile north of State Road 14.

This large, Kosciusko County bank barn is still in active use in a dairy operation. It has some original hand-hewn posts and beams with pegged, mortise and tenon construction. It has twin wagon doors centered in the side. The barn, estimated to have been built around 1860, is located on the south side of County Road 1150S about 0.2 mile east of County Road 700E.

This bank barn with a Dutch gambrel roof and twin off-center wagon doors is estimated to have been built in the late 1800s and is still in active use. The original wood-louvered vents have been replaced by windows, and the shed on the gable end appears to be a later addition. This barn originally had a fieldstone foundation. Its roof and sides have recently been covered with aluminum. The barn is located on the west side of County Road 13 about 0.3 mile south of County Road 50 in Elkhart County.

Below: This Elkhart County barn is white, neat, and well maintained which is typical of many "Plain People" barns. The original barn was a standard gable-roofed barn. The section with the Dutch gambrel roof was added later. The original fieldstone foundation has been partially replaced and the newer addition has a concrete block foundation. The barn is located on the north side of County Road 54 about 0.3 mile south of County Road 50.

This bank barn with a full forbay shows the "Plain People" influence in barn construction. The original fieldstone foundation has been partially replaced with concrete blocks. Note the twin doors from the threshing floor in the forbay. This barn is situated on the west side of County Road 25 just south of U.S. 6 in Elkhart County.

This 1890 St. Joseph County bank barn sits on the west side of State Road 23 about one mile south of the U.S. 20–31 bypass south of South Bend. It has the original fieldstone foundation, centered-side wagon door, and gable roof. It has hand-hewn beams and posts with pegged mortise and tenon joints. It was recently painted and its owner is just as colorful as the barn. The barn is no longer used in an agricultural operation.

This old, modified bank barn is no longer used in an agricultural operation but has been preserved by its urban owner. The repaired fieldstone foundation and vertical plank siding appear similar to when the barn was built in the mid-1800s. The wagon door was located on the gable end. It appears that there was no exterior hay door. The barn is located east of Mishawaka on the south side of Jackson Road just east of Fir Road in St. Joseph County.

This horse barn is a modern pole barn which is much cheaper to build than the original Indiana barns. The complex is located on the east side of State Road 331 just south of Wyatt in St. Joseph County.

Names are frequently designed on barn roofs. This barn is on the northeast corner of U.S. 31 and State Road 16 in Miami County.

Tractor design in roof. Barn is located on the west side of U.S. 31 about 0.3 mile north of County Road 800N, Miami County.

American flag and eagle design on roof of this Miami County barn located on the west side of U.S. 31 about 0.4 mile south of State Road 16.

Barn Roof Designs

Horses designed in roof shingles. Barn is located on west side of State Road 13 at the north edge of Manchester, Wabash County.

Left: Doud Orchards on east side of State Road 19 about 0.5 mile north of County Road 850N, Miami County.

Below: This Miami County round barn has several animals designed into the roof. It is located on the west side of old U.S. 31 about 0.8 mile south of County Road 1350N. This barn is further documented in Hanou, A Round Indiana, *1993.*

Below: LaPorte County is the home of these rare, glazed tile, connected, twin silos.

Right: Although rare in other parts of the state, a few brick silos remain in Clinton County.

Above: This LaPorte County barn is estimated to have been built in the 1850s. It has part of the original fieldstone foundation, wood-louvered vents, wagon door centered in the side, the remains of the wood vent above the wagon door, plank and batten siding, and the posts and beams are notched logs. The barn appears to be abandoned. It is located on the south side of U.S. 6 at the LaPorte-Porter County Line.

Top Right: This Starke County raised barn has a Dutch gambrel roof and wagon door in the gable end. The barn siding has been covered with composition sheeting and in the process the gable-end hay door was covered over, but the extended roof ridge that served as a hay hood remains. It is estimated that the barn was built in the 1930s and was originally used as a dairy barn. The barn is no longer a working barn. It is located on the east side of County Road 250W about 0.5 mile north of State Road 10-39.

Right: This glazed-tile block barn was originally used as a dairy barn. It was heavily damaged by a tornado in the 1960s, but was repaired to its original state. Now used for machinery storage, it is located on the north side of County Road 1150S about 0.3 mile east of State Road 39 in LaPorte County.

Barns in Northwest Indiana

Several counties in the North west Region were originally wet, low-lying areas. When the Kankakee marshes were drained, it opened the highly productive soil to crop production. Sandy soils predominate along the Lake Michigan areas. Dairy production occurs in the northern areas while cattle, hog, and crop production occur in the southern portion of the region. The northwestern portion is highly urbanized and industrialized.

This dairy barn has a Dutch gambrel roof, hooded hay door, and gable-end wagon door. The siding is vertical plank with batten strips. The estimated time of construction is the early 1900s. It is located on the west side of State Road 2 at County Road 600S in Porter County.

Left: This old, brick, Porter County bank barn is unique. It has hay doors in the side as well as the gable end, a wood cupola with louvered vents, and a gable dormer. The wood portion to the right of the photo may have been added later, but it is also quite old and has vertical planking siding with batten strips. The barn has been remodeled and is still a working cattle barn. The barn, estimated to have been built during the mid- to late-1800s, is located on the west side of State Road 2 about 0.3 mile south of County Road 150W.

Right: This is a large, raised, working barn with the first floor formed by manufactured concrete blocks that have been shaped to resemble cut limestone blocks. The wagon door is in the gable end and the hay door is not shown. The barn has a Dutch gambrel roof. The barn, estimated to have been built in the 1900 to 1920 era, is located on the east side of State Road 2 about 0.2 mile north of County Road 450S in Porter County.

The date of 1893 painted on the side is the approximate date the barn was built. It has a gambrel roof and centered side wagon door. The siding is vertical planks with batten strips. It is estimated that the sheds on the right and left were added after the original barn was built. The barn is located on the north side of State Road 2 just east of State Road 55, Lake County.

It was not determined that the date of 1875 painted on the barn was the actual date this Newton County barn was built. More likely, it was the date the farm was established. This working barn has a fully enclosed hay hood, centered side wagon door and gable roof. The barn has been remodeled and has been well maintained. It is located on the south side of County Road 1200S just west of the Newton County Fairgrounds.

This older standard barn with a gable roof, wooden cupola, and gable-end wagon door is unusual with gabled porches over the two side doors. The wood-louvered vents in the cupola and barn sides have been boarded over. It is estimated that this barn was built around the 1870s. It is located on the east side of State Road 55 about 0.3 mile north of County Road 700S in Newton County.

This Newton County standard barn sits at the southeast edge of Morroco. The gable roof has gable-roofed dormers on the sides, a wooden cupola, and an extended roof ridge that serves as a hay hood for the gable-end hay door. The wagon door is centered on the side (not shown). The original wood-louvered vents have been replaced or boarded over. The barn, estimated to have been built during the late 1800s, sits on the west side of County Road 275W about 0.3 mile north of State Road 114.

This Newton County working cattle barn is located on the north side of County Road 1400S about 0.2 mile east of County Road 525W. It appears that sections of this barn were built at different times. The back portion is of a raised barn construction with the first floor wall being fieldstones. The right-hand portion has a cut fieldstone foundation. The dates these sections were built was not determined, but it would appear that possible times would be late 1800s.

This large, Jasper County barn was razed in December 1996. When I-65 from Indianapolis to Chicago was constructed, it divided the farm and changed the way the farm was operated. The barn had a gable roof, hay door, hay hood, and twin cupolas with wood-louvered vents. The stately barn, estimated to have been built in the late 1800s, still had character in spite of the loss of utilitarian value to the farmer. The barn was located on the north side of State Road 114 at I-65.

This immaculately maintained barn is an example of the entire farmstead. The barn was originally a dairy barn, then modified for hog production, and now is used for storage. It has a gambrel roof with a hay hood over the hay door. The wagon door is centered on the side (not shown). The barn sits on the north side of County Road 1000S just east of County Road 880W in Jasper County.

This drive-through corncrib has a cupola that houses an enclosed grain leg. The cupola has a gable roof while the crib has a Dutch gambrel roof. The crib is located on the east side of County Road 400W about 0.5 mile north of County Road 400S in Jasper County.

This well-maintained Pulaski County barn was built in 1930. It has a monitor roof with a hay door and hay hood in the gable end. The wagon door is in the gable end (not shown). Note the 1950s, steel-wheeled, side-delivery rake in front of the barn. The barn is on the east side of County Road 1425W about 0.5 mile north of State Road 14.

Above: This fieldstone and brick barn in White County is very unique. The locals state there were two other similar barns that were part of a large dairy. This is the only remaining one. The barn has a gambrel roof, four gable-dormers on each side, a slate roof, and an unusual vent in the gable ends. Unfortunately, the date the barns were built was not determined. It is estimated to be around the late 1800s to early 1900s. The wagon door is on the gable end. The barn is on the corner of Airport and Freeman Roads just south of Monticello.

Above: Details of the dormers, brickwork around the windows and doors, and the slate roof.

Right: Details of the chain lock for the large wagon door.

This Benton County barn is attractive with its weathered colors. It is a standard barn with a fieldstone foundation, a gable roof, twin cupolas, and a wagon door on the gable end. The vents on the cupolas, side, and gable ends were originally protected by wood louvers. Part of the siding is vertical planks with batten strips. The barn, estimated to have been built in the late 1800s, is located on the west side of County Road 900W about 0.1 mile north of County Road 500N.

This late 1800s Benton County barn has been immaculately maintained over the years. It was originally a cattle barn and is still a working barn. It is a standard barn with a gable roof and an unusual eight-sided cupola. The wagon doors are on either side of the gable end. The hay door is protected by a hay hood. Most of the wood-louvered vents on the cupola, sides, and gable ends remain intact. The original foundation has been replaced by poured concrete. The barn is located on the west side of County Road 600N at 2064N 600W.

This "Lonesome Crib," in the author's terms, is typical of numerous Benton County corncribs. They frequently stand alone with no other buildings close by, especially in the prairie areas. Many have a windmill nearby. This corncrib is located in the middle of a field on the west side of County Road 1000W about 0.5 mile north of County Road 200S.

This unique barn was originally built for feeding cattle. The tall portion to the right with the hip roof houses twin silos with the feeding shed on the left. The siding remains primarily vertical plank with batten strips. Date of construction is unknown to the author. The Benton County barn is north of the barn on the preceding page on the west side of County Road 600N.

This three-gabled barn, although old and unpainted, still has considerable character and continues as a working barn. The barn is just north of Independence, a historic early settlement in the area. The barn has hand-hewn log posts and beams, a fieldstone foundation, siding of vertical planks with batten strips, hay hood over the hay door, and a broken gable roof on one side. This barn, estimated to have been built in the 1850s to 1870s, is located on the east side of the Independence-Pine Village Road about 0.1 mile north of the historic Independence Cemetery in Warren County.

Top Right: This well-constructed barn was built in 1934 as a horse barn and continued that use until the farmer passed away a few years ago. The stalls, feeding bunks, door hinges, latches, and superb neatness of the barn are as he left them. Today, the barn is a symbol of the ingenuity and craftsmanship of Indiana farmers. The barn has a Dutch gambrel roof, poured concrete foundation, wagon doors on both gable ends, and hay door with hay hood (not shown). The barn is located on the westerly side of Akers Road about 0.3 mile north of County Road 200E in Warren County.

Barns in West Central Indiana

The topography of the West Central Region is highly variable from generally level to rolling. The southern portions of the region are hilly and heavily timbered. Crop production is the major agricultural enterprise. In Tippecanoe County agricultural land is being rapidly taken out of production and developed for housing, industry, shopping malls, and roads.

This Tippecanoe County barn is still a working barn. It has a gable roof, wagon door in center of side, hay door with hay hood in the gable end, sawed posts and beams, cupola with wood-louvered vents, and originally a fieldstone foundation. The barn was built around 1886 and is located on the west side of County Road 50W just north of County Road 500N.

This Tippecanoe County bank barn was built in 1888 according to a carving on one of the boards in the barn. It has a modified salt-box roof, wagon door centered on the side, a full forbay on the back side, a fieldstone foundation that has been partially replaced, and, interestingly, no large external hay door. The wood-louvered vents have been replaced by windows. The barn is located on the east side of County Road 1050E and north of County Road 300S.

This unique, modified standard barn has several features not observed in other Indiana barns. It has a Dutch gambrel roof, twin cupolas, numerous wood-louvered vents in sides and gable ends, porches over wagon doors on the sides and gable ends, and smaller hay doors on both the sides and gable ends. The cupolas have unusual steeple-style roofs with tall, metal, ball-tipped lightning rods. Estimated to have been built in the late 1800s, it is located south of Lafayette in Tippecanoe County.

The back side of the bank barn on the previous page shows the attachment of a posted feeding barn that is now used for equipment storage.

These round-roofed barns were part of the Purdue University Animal Farm until a few years ago. They were brick and frame construction and the two barns were connected. These barns and their silos were razed in December 1996 to make way for the new Purdue University Turfgrass Research Center. Photo courtesy of Arlene Blessing.

These barns and their silos were also part of the Purdue University Animal Farm until a few years ago. These barns have been remodeled, and will become part of either the West Lafayette Park System or will become a part of a housing development. They are located at the northern edge of West Lafayette in Tippecanoe County.

Left: Seen here before renovation, this Tippecanoe County barn was in a deteriorating condition and missing a cupola. The barn is located on the east side of U.S. 231 about 0.3 mile north of County Road 900N.

The same barn is shown here after renovation in 1995. Originally there were two cupolas, now one. The hay door and extended roof ridge for a hay hood have been removed. The broken gable roof and several other features have been retained.

This Montgomery County bank barn has a broken gable roof and wagon door centered on the side. It is a working barn and the windmill is operational. The barn, estimated to have been built around the early 1900s, is located on the corner of County Roads 900N and 600E.

This Montgomery County working barn is located on the west side of County Road 150E just north of County Road 650N. It has a Dutch gambrel roof, wagon door in gable end, hay door, and enclosed hay hood. Note, the hay door is no longer operational with the addition of the shed structure. The barn, estimated to have been built around 1930, has been modified for hog production.

Left: This raised barn has a brick first floor with a gable roof and wagon doors on both sides. The shed on the back appears to be a later addition. The barn, built around 1920, is located on the north side of County Road 110S about 0.4 mile west of County Road 300W in Fountain County.

Right: This standard barn has a gable roof, fieldstone foundation, cupola with wood-louvered vents, hay hood, and a counter-weighted, sliding hay door. Rather than being hinged, as most hay doors, this one slides up and down like a double-hung window. The barn is located on the north side of River Road about 0.5 mile east of County Road 325E in Fountain County.

Left: This type of concrete block corncrib was popular in Indiana, especially the west central part, in the 1930s when corn was still harvested on the ear. Several of these cribs have been modified to handle shelled grain. This crib is located on the east side of County Road 250W about 0.8 mile south of County Road 350N in Fountain County.

Right: This barn was built in 1933 on the foundation of a previous barn that burned. It was originally a horse barn, but was converted to a dairy barn in the 1940s. A silo is completely enclosed within the barn beneath the left gable end. It is a standard barn with a Dutch gambrel roof, a shed dormer on either side of the roof, hay doors and hay hoods on both gable ends, and gable-end wagon doors. The hay rope originally was pulled from the center of the building and hay could be loaded into the loft from either end. The barn continues to be a working barn, but mainly for hay storage. The barn is located on the northwest corner of State Road 63 and County Road 1000S in Vermillion County.

Left: This neat Vermillion County farmstead shows the comparison of an older barn with more modern storage facilities. The farm was established in 1824, but the barn was built possibly in the late 1800s to early 1900s. The shed on the left was a later addition. It has a gable roof, hay door on the gable end, wagon door on the side, and vertical plank and batten siding. The farmstead is located on the west side of County Road 200E and south of County Road 1450N.

This Vermillion County dairy barn is no longer a working barn. It was built in 1936, and is a Midwest Dairy Barn design. The lower portion is glazed tile. It has a Dutch gambrel roof, two shed dormers on either side of the roof, and hay door and hay hood in the gable end. There is no wagon door, as such. It is located on the east side of County Road 100W about 0.5 mile north of County Road 1350N.

As one moves into the more southern portions of the state, many barns are unpainted. This working barn, in a scenic Parke County setting, has seen better days, but is still used as a cattle barn. Date of construction is unknown to the author. It is located on the north side of State Road 59 about 0.8 mile west of County Road 800E.

This bank barn is still used to shield horses, but little else. The barn has a partial fieldstone foundation, sawed timber posts and beams, hay hood and hay door in gable end, and small wagon door off-center on the gable end. It is estimated to have been built in the late 1800s or early 1900s. It is located on the west side of County Road 625W and just north of the Parke-Vigo County line in Parke County.

Center: This is a relatively small standard barn with a salt-box roof. It is located on the north side of County Road 1200N about 0.6 mile east of County Road 500E in Putnam County. It has only a small hay door and wagon door, suggesting the barn was built around 1940.

Left: This monitor-roof barn-like structure was built recently and is used for non-agricultural purposes. It is located on the south side of County Road 725S about 0.5 mile west of County Road 625W in Putnam County.

This large, Owen County bank barn was originally painted with the orangish "turkey red" paint. The barn has a gable roof, cut fieldstone foundation, huge gable-end hay door with partially enclosed hay hood, and wagon door centered on the side (not shown). The estimated building date is in the late 1800s. The barn is no longer a working barn. It is located on the south side of State Road 46 about one mile east of Spencer.

This standard barn with a monitor roof is nestled in the hills and woods of Owen County. It has a hay hood and hay door on the gable end. The first floor on the gable end is posted and open. The barn is located on the west side of U.S. 231, about 0.7 mile north of Spencer.

This standard barn has its hay door and off-center wagon door on the side. The gable end has vertical plank and batten strip siding while the rest of the barn siding is horizontal. The barn, estimated to have been constructed around 1910 to 1920, is located south of Riley on the northeast corner of Louisville Road and Woodsmall Drive in Vigo County.

This beautiful, well-maintained Clay County barn is unique in the state. It is a standard barn with a gable roof, unusual cupola, cut-stone foundation, and twin hay and wagon doors on the gable end. The entire gable roof was extended several feet to provide protection for the doors. The wood-louvered vents on the cupola remain. Other louvered vents have been replaced by windows. The barn is located on the south side of U.S. 40 at the Clay-Putnam County line.

The McCray-Wagner barn in 1996 after moving and restoration. Photo by Jacquelyn Scott.

The McCray-Wagner Barn
(Governor McCray's Pavilion)

The savior of this historic and magnificent structure, now known as "Governor McCray's Pavilion," is John Wagner of Lafayette. The effects of time and elements had taken their toll on the barn (pavilion) over the years and the structure had only sentimental value to its owner, Henry Coussens. The roof had severely deteriorated and the entire structure would collapse in a year or two if the roof was not repaired. Repair of the roof alone was an estimated $40,000. Consequently, Mr. Coussens felt his only option was to bulldoze the structure. Dr. Wagner knew of the barn, admired its history, and what it had been so many years ago. He also needed a structure for his 300-acre Angus cattle farm. So, a deal was struck between Dr. Wagner and Mr. Coussens. Mr. Coussens sold the structure to Dr. Wagner for $1 with the understanding that the structure would be moved. A win-win situation for both men. The only problem was, Dr. Wagner's farm was twenty-two miles away in White County.

The pavilion was originally one of a dozen gleaming white farm buildings on the 1,500-acre Orchard Lake Stock farm owned by Warren T. McCray (1865–1938). The farm was about six miles northeast of Kentland in Newton County. It had pastures, fields, an apple orchard, and a herd of five hundred purebred Hereford cattle. Mr. McCray was a Kentland banker, land speculator, cattleman, and Indiana's twenty-ninth governor. His prized bull, Perfection Fairfax, was the Grand Champion Bull at the 1907 International Livestock Exhibition in Scotland.

The pavilion is thought to have been built in 1896. It was, and still is, an eight-sided building eighty feet by forty feet by fifty-four feet high. It had an octagon-shaped cupola with large windows, a hip on gable roof, and a total of six shed dormers. Wagon doors were located on each side and on both ends. A large cattle holding barn was attached on the back of the pavilion. Starting in 1908, Governor McCray held his breeding livestock auctions in the pavilion. His auctions were international events. Cattle buyers from Europe, South America, Canada, as well as all parts of the United States attended. The auctions reportedly attracted up to five thousand people for each event. Inside the pavilion were bleacher seats that ringed the sale floor, with the raised auctioneer's platform on the back side of the ring. The bleachers would accommodate up to eighteen hundred cattle buyers. In one auction, 125 cattle were sold for $436,250, with some animals bringing up to $20,000 each (Wagner, 1996). Auctions were held in the pavilion until 1938 when Governor McCray died unexpectedly. The last auction was April 14, 1938. After his death, the herd was sold, and the pavilion sat unused for many years.

Dr. Wagner began dismantling the pavilion on April 19, 1996. Each board, post, beam, brace, and window was numbered so the pavilion could be reconstructed in exactly the same manner. The carefully removed parts were loaded onto semi-trailers and transported twenty-two miles to his White County farm. Amos Schwartz of Geneva, Indiana, was hired as contractor to reassemble the pavilion. Portions of the pavilion that were badly deteriorated or missing were replaced by custom milling replacement parts based on old records and photographs. The red on white color of the original structure was carefully reproduced. The faithful and exacting reconstruction of the pavilion was completed in November 1996. It is as close to the original as humanly possible, except the severely deteriorated cattle holding shed attached on the rear of the pavilion was not moved and the roof sheathing as well as the roof shingles are new. The cattle holding shed has been replaced with an aluminum-clad pole barn. Dr. Wagner stated that the $80,000 cost to replace the original tongue and grove roof sheathing was too much, so he used plywood sheathing instead. The color of the composition shingles was matched as closely as possible to that of the original wood shingles.

Dr. Wagner's first livestock auction held in the reconstructed pavilion was Saturday, November 23, 1996. In honor of Governor McCray's bull "Perfection Fairfax," Dr. Wagner named his prized Angus bull "Return to Perfection." Today the magnificent and faithfully reconstructed pavilion sits on the south side of County Road 600S about 0.5 mile west of County Road 1000W in White County.

Dr. Wagner stated, "It probably would have been cheaper to build a new structure, but you can't build history, feeling, and warmth. If this old barn could talk, we could spend hours listening."

Dr. Wagner has named the pavilion, "Governor McCray's Pavilion." The author, however, has taken the liberty to call the pavilion, at least in this book, "The McCray-Wagner Barn" to honor Dr. Wagner's dedication, perseverance, time, vision and expense to preserve a part of Indiana's agricultural history. A job well done!

The McCray-Wagner barn setting. Barns to the right and rear are not part of the original barn.

This 1912 photograph shows the original sale pavilion at Orchard Lake Stock Farm and some of the five thousand people that attended that sale. Photo printed by permission of John Wagner.

This 1919 photo shows three of the dozen buildings at the Orchard Lake Stock Farm. The pavilion is the middle structure. Photo printed by permission of John Wagner.

This standard barn has a gable roof, cupola, hay door, partially enclosed hay hood, and wagon door on the gable end. The original wood-louvered vents have been covered over. The Sullivan County barn, estimated to have been built in the early 1900s, is located on the east side of U.S. 41 at the south edge of Sullivan.

Top Right: This attractive Sullivan County barn is unusual in that the sills sit atop concrete pillars rather than on a continuous foundation. It is a standard barn with a gable roof, wagon door in the gable end, hay door, and fully enclosed hay hood. The estimated building date is the late 1800s to early 1900. The barn is located on the north side of County Road 500N and just east of State Road 63.

Right: This barn was once a livery stable in Merom. It is located near Merom Bluff Park, the site of ten-day Merom Bluff Chautauquas from 1905 to 1936. The barn has not been used as a livery for many years, but stands as a monument to those earlier days when Carrie Nation, William Jennings Bryan, William Howard Taft, Warren Harding, Billy Sunday, and others spoke in the park. The livery is located on the corner of Market and Second Streets in Merom, Sullivan County.

Barns in Southwest Indiana

The topography of the Southwest Region varies from level bottom lands to rolling uplands. Portions of the region have or are underlain with coal, and mining occurs. Most mines are strip mines. Many barns originally in these regions were razed or moved to make way for the strip mines. Hilly, wooded, and picturesque areas occur in the eastern portion. The southwestern portion is highly productive for crops and livestock. Tobacco production occurs on the upland soils in the southern portion. Portions of Knox and Vanderburgh Counties are increasingly urbanized and industrialized.

This abandoned, drive-through, corncrib is unusual in that the siding is composed of slats on a 45-degree angle rather than vertical or horizontal. The roof sheathing on the right side is now missing giving the shadow effect on the crib. The date this crib was built is unknown. It is located on the east side of State Road 59 about 0.5 mile north of the Greene-Knox County line in Greene County.

This Martin County barn has both hand-hewn and sawed timbers. It has a gable roof, cupola with wood-louvered vents, wagon door in gable end to left (not shown), hay door, and hay hood. The barn originally had a cut-stone foundation. The age of the barn is unknown, and it is little used. It is located on the east side of State Road 450 about 2.4 miles north of Dover Hill.

This Daviess County working barn is patterned after the raised barn style. It has a recent concrete block foundation, a Dutch gambrel roof, a gable-end full-pent roof, and the wagon door is off-center on the side. The estimated building date is about the 1930s. It is located on the north side of County Road 100S about 0.1 mile east of County Road 1175 E.

This older barn has an unusual design with what appears to be a hip on gable roof. It is not clear if this design was the result of sheds added on the side and left gable, or if the sheds were part of the original design. The rest of the roof is a gable with hay door, hay hood, and wagon door in the gable end. The estimated building date is about the early 1900s. It has recently been covered with aluminum siding. The barn is located at the junction of Old U.S. 50 and U.S. 50-150 in Daviess County.

A hip on gable roof barn that originally had a fieldstone pillar foundation. The wagon door is on the side (not shown), and the barn appears to be little used. It is located on the north side of County Road 300S about 0.2 mile west of State Road 57 in Daviess County.

This true round barn with wood-louvered vents in the cupola was built in 1908. It has vertical plank and batten strip siding. This barn is also mentioned by John T. Hanou in A Round Indiana, *1993. The barn is located on the south side of County Road 450S and just west of State Road 57 in Daviess County.*

All barns were the pride and joy of the farmer who built them. This is one of those barns that lost its utilitarian value, became too expensive to maintain, and is now called "Yesterday's Dreams." Although abandoned, the barn still has character. It is located west of U.S. 41 in Knox County.

This well-maintained barn was moved to this location in 1938. The date of its original construction is unknown. It may have been moved to make way for strip mines from a more northern county, but that is unconfirmed. The barn was originally a horse barn, but is now a storage building. It has a broken gable roof, hay door and hay hood in the gable end (not shown), and a posted pent roof. Note the rooster weather vane. The barn is located on the south side of State Road 64 about 0.2 mile west of County Road 950E in Gibson County.

This barn, built in 1930, has a Dutch gambrel roof and an off-center wagon door on the side. The hay hood remains, but the hay door has been covered over with aluminum siding. The barn is located on the south side of State Road 64 just east of County Road 850E in Gibson County.

This Gibson County barn was built in 1880, and was originally painted black. It still retains some of the black color. It has hand-hewn timber posts and beams with pegged, mortise and tenon joints. The original foundation was fieldstone. The wagon door is on the side. The hay door and hay hood are on the gable end to the right (not shown). The barn is located on the north side of County Road 250S about 0.1 mile east of County Road 850E.

This all but abandoned barn is still in reasonable condition. It has a gable roof, off-center wagon door on the side, hay door and hay hood in the gable end. The date of construction is unknown. The barn is located on the southerly side of County Road 900E about 0.1 mile west of State Road 257 in Pike County.

This monitor-roofed barn with the smokestacks of Petersburg Power Station in the background typifies Pike County. The county is largely underlain with coal and many of the barns that originally stood were razed or moved to make way for the strip mines. This barn is located on the west side of State Road 57 at County Road 650N.

Typical of many southern Indiana barns, this one has never been painted. It is a standard barn with a gable roof, fieldstone foundation, and hand-hewn timbers with pegged, mortise and tenon joints. The hay door, hay hood, and wagon door are on the gable end (not shown). The estimated date of construction is around the late 1800s. The barn is located on the southeast corner of County Road 650W and Duff Road in Dubois County.

This Spencer County dairy barn is of concrete block and wood construction, possibly built in the 1930s to early 1940s. The twin wagon doors are off-center on the side and the hay door is in the left gable (not shown). It has a Dutch gambrel roof, and would classify as a standard raised barn. The barn, no longer used as an active dairy barn, is located on the west side of County Road 350W about 0.2 mile south of State Road 66.

Below: Since air circulation is imperative for tobacco curing, this weathered corner of the barn has not been repaired.

Above: The type of large tobacco barns found in Kentucky are difficult to find in Indiana. In most cases, Indiana tobacco farmers have modified multipurpose barns such as this to cure their tobacco. This is an old raised barn with a recent concrete block foundation. It has a gable roof, but little else can be determined due to the various modifications. The barn is located on the south side of County Road 200N about 0.1 mile east of County Road 700W in Spencer County.

This well-maintained 1890 barn has hand-hewn timber beams and posts with pegged mortise and tenon joints. The plank and batten strip siding remains intact. It has a fieldstone foundation, gable roof, twin wagon doors off-center on the side, and hay door in the gable end. The barn is located at 10655 Old Boonville Highway in Warrick County.

Right: These unusual twin barns were once part of a large dairy operation that served the Evansville area. They are brick and frame with modified gambrel roofs. The hay doors are on the gable ends. The barn on the left has a large glass-block window to the left of the small door. The purpose of the unusual gable-end extensions is unknown. The barns are located on the west side of Old Boonville Highway about 0.5 mile north of State Road 62 in Warrick County.

This Warrick County barn is thought to have been built in the late 1800s. The cupola, originally had wood-louvered vents and there are indications of the large wood-louvered vent above the wagon door. The original wagon door has been replaced with a larger, modern door. The barn has a gable roof with the hay door and enclosed hay hood in the gable end. This working barn is located on the south side of County Road 550S about 0.2 mile east of State Road 61.

This is one of very few barns in the state that has a full basement large enough to accommodate small horses. The original barn, estimated to have been built in the early 1900s, had a Dutch gambrel roof and the wagon door centered on the side. The hip-roof design was the result of adding sheds on the gable end and left side at a later date. The barn is located just to the east of U.S. 41 on Inglefield Road in Vanderburgh County.

This standard barn has a broken-gable roof, wagon door centered on the side, and concrete block foundation. The barn, built on an unknown date, is located on the north side of Base Line Road at Cemetery Road in Vanderburgh County.

This is a three-bay barn with the center bay housing a corncrib (the crib portion can barely be seen in the Fig.). The gable-roofed barn had a hay door and a hay hood on the gable end. The date of construction is unknown, but estimated to be the late 1800s. The barn is located on the west side of State Road 69 south of Mt. Vernon in Posey County.

This Posey County standard barn is located on the west side of State Road 69 across the road from Hovey Lake. It has a gable roof with hay hood, hay door, and wagon door on the gable end. The estimated date the barn was built is the early 1900s.

This hip on gable-roof barn also has a gable dormer-like roof offering some protection to the wagon door on the side. The date the barn was built is unknown. It is located on the north side of County Road 850S about 0.75 mile west of County Road 500E in Posey County.

A reconstructed early settler log barn. Notice the notched logs, the wood-shingle roof, and the overlapped shingles at the roof ridge. The purpose of this type of barn was to house food and grain for the farm family. In the early days, farmers did not have animals as they relied on wild game for their meat. This Posey County building is in New Harmony at the reconstructed village.

This drive-through crib for ear corn is 120 feet long with numerous interior and exterior doors for loading and unloading corn. It sits on a fieldstone foundation. While the exact date the crib was built is not known, it reportedly was around the 1880s. The crib, no longer used, is located on the east side of State Road 69 north of Hovey Lake in Posey County.

Left: This small building, known as a "gear shed," was the barn used to house the mule harnesses and gear. The posted overhangs were for protection of wagons and other equipment. The barn, which originally stood on fieldstone piers, is located on the same farm as the drive-through crib. A windmill still stands in the background.

Bottom Left: This Posey County barn was known as the "milk barn" and was used to milk the two or three cows kept on the farm back in the mid- to late-1880s. This barn is on the same farm as the drive-through crib and gear shed. These buildings have been maintained as remembrances by the farm owners, even though they do not farm and the buildings are little used.

One of the older Posey County barns still standing. It was built in the late 1800s with hand-hewn beams and posts with pegged mortise and tenon joints. The original cypress siding is covered with sheet metal. It has a cupola with wood-louvered vents. The hay door and hay hood are in the gable end. Still a working barn, it is located at 9315 Ries Road in Posey County. Photo by David Ries.

This stilt barn, located in Point Township, Posey County, was built in the early 1930s. It was at the very tip of Indiana within 0.5 mile of the Wabash and Ohio Rivers. David Ries stated that the area flooded every year, thus the ten-foot high stilts that supported the barn. The stilts also extended six feet below ground. Concrete for the stilts was hand-mixed one bucket at a time and dumped into the stilt forms. Mr. Ries also stated, "By taking off the rear tire of a Model A Ford, they blocked up the rear end and used the wheel as a winch to lift the concrete and lumber. At the end of the day, they would put the tire back on and drive the sixteen miles back to Mt. Vernon." The Barn was a slotted corncrib with a center alley for unloading ear corn to a wagon or truck, or a barge if the water was high enough. Until around 1990, this was the only barn in the flood plain that withstood the flood water and winds of time. The barn was destroyed by fire at that time. Photo by David Ries.

This Grant County hay hood is formed by an extension of the roof ridge above the hay door. The barn is located on the west side of County Road 1150E about 1/3 mile north of County Road 600S.

Right: This combination of a fully enclosed hay hood with the lower portion of the hood also serving as the hay door is on a Sullivan County barn on the east side of State Road 63 at County Road 500N.

Hay Hoods

This partially enclosed hay hood is on a Clay County barn located on the south side of State Road 46 about 2 miles east of Bowling Green.

This Jefferson County barn has a fully enclosed hay hood. Other photographs of this barn are on page 50.

This old hay hook for loose-hay with remnents of hay rope is hanging in the ridge of a Tippecanoe County barn. The barn was razed in 1992.

Right: A unique type of hay hood is formed by the extension of the entire gable roof several feet. Note the hay hook hanging form the gable end. This is a later style hay hook used primarily for baled hay, not loose hay. Another photo of this Clay County barn is on page 129.

Glossary

Amish Barn: See **"Plain People" Barn.**

Bank Barn: A barn usually built in the side of a hill, with wagon door entrance to the second floor known as the threshing floor. The first floor opened at ground level usually on the opposite side of the threshing floor entrance, which may be on the side or gable end.

Cupola: A relatively small wooden structure at the ridge of a barn roof, usually near the center of the roof. Cupolas have their own roofs which may or may not be of the same slope or design as the barn roof. Cupolas have wood-louvered vents or windows for lighting and ventilation. In general, cupolas were replaced by metal roof vents on barns built after about 1910 to 1920.

Cut-stone Foundation: The foundation of a barn made from fieldstone or quarried stones that have been partially or totally shaped to closely fit, usually by hammer and chisels. Cut-stone foundations may be of fieldstones, limestone, or sandstone.

Dormer: A vertical projection built out from a sloping roof with a window or louvers.

Extended Barn: Side additions made to the main barn structure usually at the time the barn was built. The slope of the roof may be continuous to the eaves or may change pitch at the additions resulting in a roof described as a broken gable. It is often difficult to determine if the additions were made at the time the barn was built or added at a later date.

Feeder Barn: A barn with long, gently sloping roofs that had eaves close to the ground.

Fieldstone Foundation: Rocks or boulders of various sizes collected from fields in glaciated areas and used in building a foundation. The foundation is generally continuous under the perimeter of the barn as differentiated from a pier foundation.

Forbay: A cantilevered extension or overhang of the threshing floor (second floor) beyond the first floor and foundation wall. Forbays are usually on the down-slope side of bank barns and opposite the wagon doors.

Gable End: An end wall bearing a gable.

Gable Roof: A roof sloping downward in two parts at an angle from a central ridge, so as to leave a gable at each end.

Gambrel Roof: A gable roof, each side of which has a shallower slope above a steeper one.

Hay Hood: Offered protection against the elements for the hay door. Hay hoods were sometimes just an extension of the roof ridge. Others were partially or entirely enclosed offering more protection.

Meadow Barn: A small building that usually sat alone in a field. Usually built there for convenience of storing hay or grain close to where it was produced, eliminating long trips back to the main barns. They were also used for housing animals in the fields.

Midwest Dairy Barn, also called **Wisconsin Dairy Barn:** A barn built for dairy cattle. Most were designed by various Agricultural Experiment Stations and the plans were purchased by farmers.

Modified Bank Barn: A barn constructed on relatively level terrain rather the being built into a hill. An incline drive leads to the second floor (threshing floor). Wagon doors may be on either sides or gable ends.

Monitor Roof: A raised portion of a barn at the roof ridge used to provide wood-louvered vents or windows for lighting and ventilation. The monitor roof has its own roof generally with the same slope as the rest of the barn roof.

Pent Roof, also called a **Shed Roof:** A roof with a single slope; a sloping roof projecting from a wall or the side of a building.

Pier Foundation, also called a **Pillar Foundation:** An upright discontinuous foundation of fieldstone, cut stone, brick, or concrete that supports the sills of a structure and also permits air circulation beneath the lower building floor.

"Plain People" Barn: A barn built by Amish or a similar religious group. Most were patterned after the Pennsylvania Dutch barns. Most were bank barns, with gable roofs, with a full forbay on the opposite side of the threshing floor doors. They usually had large doors leading from the second floor for tossing down hay or straw. The forbay was also used as protection of equipment.

Raised Barn: A barn built on level ground with the sides and ends of the first level being constructed of foundation-type materials. Wagon doors are located on the first floor. There is no incline drive for entrance to the second floor as in bank or modified bank barns.

Round Barn: A circular or polygonal barn.

Shed Roof: See **Pent Roof.**

Standard Barn: A multistory rectangular barn, commonly built in the Midwest, with no distinguishing characteristics.

Three-Gable Barn, also known as a **Three-end Barn:** A barn usually in an L- or T-shaped configuration.

Threshing Floor: The second floor of a barn. See also **Bank Barn.**

Working Barn: A barn that is actively used in a farming operation.

Bibliography

Burden, Ernest. *Living Barns: How to Find and Restore a Barn of Your Own.* New York: Bonanza Books, 1984.

Caravan, Jill. *American Barns: A Pictorial History.* Philadelphia: Courage Books, 1995.

Gann, R. W. *Indiana Agricultural Statistics—1990.* West Lafayette: Indiana Agricultural Statistics Service, Purdue University, 1990.

________. *Indiana Agricultural Statistics—1994–95.* West Lafayette: Indiana Agricultural Statistics Service, Purdue University, 1995.

Gille, Frank H., Ed. *The Encyclopedia of Indiana.* 2nd ed. St. Clair Shores, Mich.: Somerset Publishers, 1976.

Hanou, John T. *A Round Indiana: Round Barns in the Hoosier State.* West Lafayette, Ind.: Purdue University Press, 1993.

Klamkin, Charles. *Barns: Their History, Preservation, and Restoration.* New York: Hawthorn Books, Inc., 1973.

Noble, Allen G. and Richard K. Cleek. *The Old Barn Book.* New Brunswick, N.J.: Rutgers University Press, 1995.

Puetz, C. J. *Indiana's County Maps.* Lydon Station, Wisc.: Thomas Publications, Ltd., n.d.

Sloane, Eric. *An Age of Barns.* New York: Funk and Wagnalls Publishing Co., 1967; 3rd Printing, New York: Ballantine Books, March 1975.

Wagner, John, Dr. Personal communications, 1996.

Index

About the Author

Donald H. Scott has been a professor of plant pathology and extension plant pathologist in the Department of Botany and Plant Pathology at Purdue University since 1968. He is a native Hoosier originally from Marion County, Indiana. Don received his B.S. degree from Purdue University in 1956. His M.S. and Ph.D. degrees in plant pathology were received from the University of Illinois in Urbana, Illinois, in 1964 and 1968, respectively.

Don Scott has had an interest in agriculture and farm barns since his early years on a Marion County farm. The smells of newly harvested hay stored in the mow, laying on the hay listening to a soft rain lightly drumming on the tin roof, and dreaming of things to come remain vivid in Don's mind. During the past twenty-nine years, as he has traveled throughout Indiana in the course of his duties for Purdue University, it was obvious that the agricultural barns that once dotted the Hoosier landscape were rapidly disappearing. Don started photographing barns around the state as a hobby when it became painfully obvious that the majority of these barns would be forever lost and forgotten within the next several years. His desire to document the existence of barns throughout the state has led to the publication of this book, *Barns of Indiana*. He has photographed and catalogued barns in all ninety-two Indiana counties, and has a collection of over seven hundred barn photographs.

He has authored or co-authored over two hundred research and extension publications in plant pathology. Scott is a member of Sigma Xi, Gamma Delta Sigma, The American Phytopathological Society, Indiana Academy of Sciences, American Soybean Association, Optimists International, and is cited in *American Men and Women of Science*, and *Who's Who in the Midwest*. He and his wife, Jackie, are the parents of four grown children and reside in West Lafayette, Indiana.

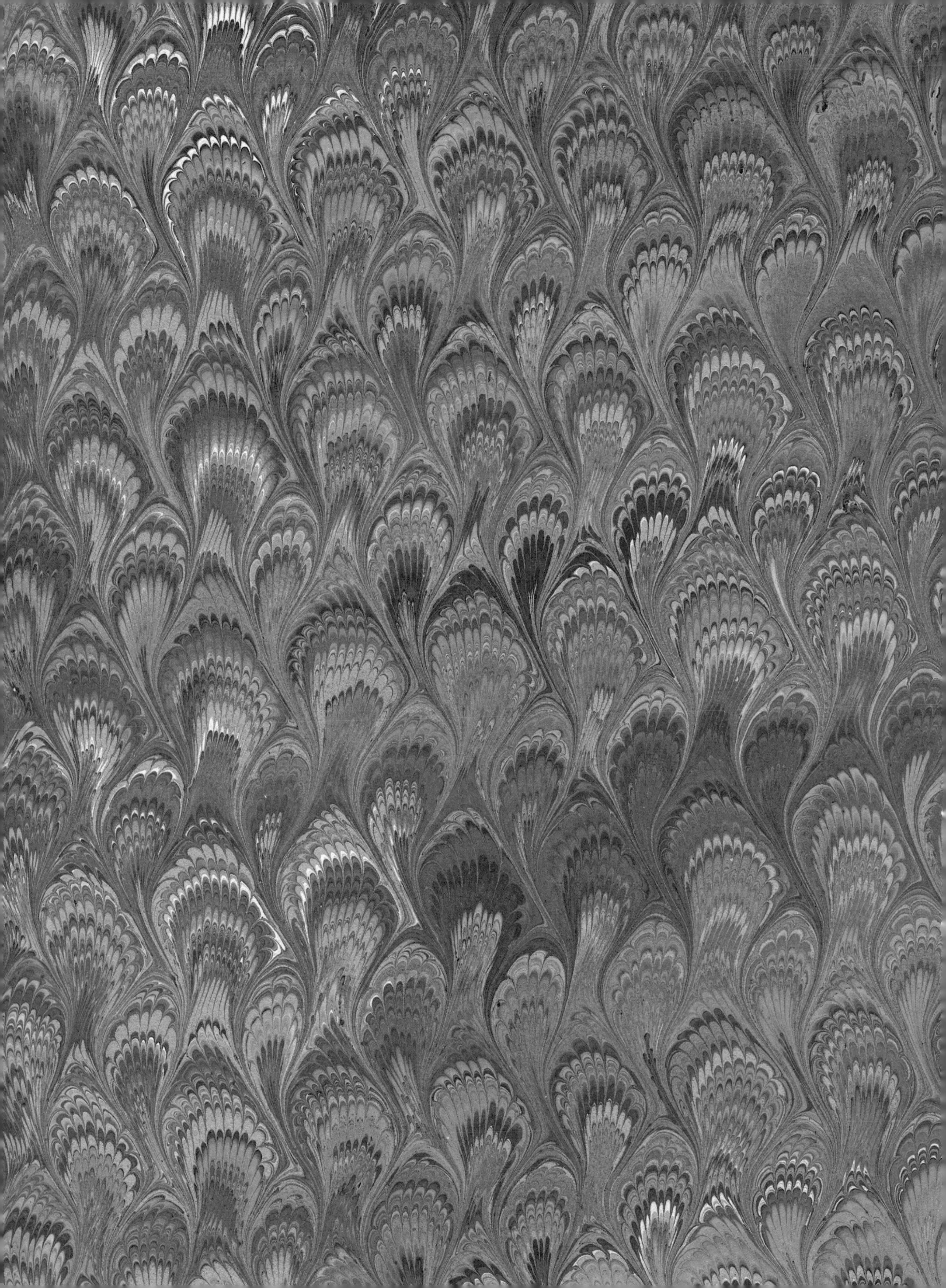